## Understanding Visuals in the Life Sciences

From photographs to micrographs, from the various types of graphs to fun, interactive visuals and games, there are many different forms in which science can be visualized. However, all of these forms of visualization in the life sciences are susceptible to misunderstandings and misinformation. This accessible and concise book demonstrates the misconceptions surrounding the visuals used in popular life science communication. Richly illustrated in color, this guide is packed with examples of commonly used visual types: photographs, micrographs, illustrations, graphs, interactive visuals, and info-graphics allowing visual creators to produce more effective visuals that aspire to being both attractive and informative for their target audience. It also encourages nonspecialist readers to be more empowered and critical, to ask difficult questions, and to cultivate true engagement with science. This book is an invaluable resource for life scientists and science communicators, and anyone who creates visuals for public or nonspecialist readers.

Han Yu is Professor of Science Communication at the Department of English at Kansas State University, and a Fellow of the Association of Teachers of Technical Writing. Her research focuses on popular science communication and visual rhetoric. She is the author of *The Curious Human Knee* (2023); *Mind Thief: The Story of Alzheimer's* (2021); *Communicating Genetics: Visualizations and Representations* (2017); and *The Other Kind of Funnies: Comics in Technical Communication* (2015).

The *Understanding Life* series is for anyone wanting an engaging and concise way into a key biological topic. Offering a multidisciplinary perspective, these accessible guides address common misconceptions and misunderstandings in a thoughtful way to help stimulate debate and encourage a more in-depth understanding. Written by leading thinkers in each field, these books are for anyone wanting an expert overview that will enable clearer thinking on each topic.

Series Editor: Kostas Kampourakis http://kampourakis.com

**Published titles:**

| | | |
|---|---|---|
| *Understanding Evolution* | Kostas Kampourakis | 9781108746083 |
| *Understanding Coronavirus* | Raul Rabadan | 9781108826716 |
| *Understanding Development* | Alessandro Minelli | 9781108799232 |
| *Understanding Evo-Devo* | Wallace Arthur | 9781108819466 |
| *Understanding Genes* | Kostas Kampourakis | 9781108812825 |
| *Understanding DNA Ancestry* | Sheldon Krimsky | 9781108816038 |
| *Understanding Intelligence* | Ken Richardson | 9781108940368 |
| *Understanding Metaphors in the Life Sciences* | Andrew S. Reynolds | 9781108940498 |
| *Understanding Cancer* | Robin Hesketh | 9781009005999 |
| *Understanding How Science Explains the World* | Kevin McCain | 9781108995504 |
| *Understanding Race* | Rob DeSalle and Ian Tattersall | 9781009055581 |
| *Understanding Human Evolution* | Ian Tattersall | 9781009101998 |
| *Understanding Human Metabolism* | Keith N. Frayn | 9781009108522 |
| *Understanding Fertility* | Gab Kovacs | 9781009054164 |
| *Understanding Forensic DNA* | Suzanne Bell and John M. Butler | 9781009044011 |
| *Understanding Natural Selection* | Michael Ruse | 9781009088329 |
| *Understanding Life in the Universe* | Wallace Arthur | 9781009207324 |
| *Understanding Species* | John S. Wilkins | 9781108987196 |

# Understanding Visuals in the Life Sciences

HAN YU
Kansas State University

Shaftesbury Road, Cambridge CB2 8EA, United Kingdom

One Liberty Plaza, 20th Floor, New York, NY 10006, USA

477 Williamstown Road, Port Melbourne, VIC 3207, Australia

314–321, 3rd Floor, Plot 3, Splendor Forum, Jasola District Centre,
New Delhi – 110025, India

103 Penang Road, #05–06/07, Visioncrest Commercial, Singapore 238467

Cambridge University Press is part of Cambridge University Press & Assessment,
a department of the University of Cambridge.

We share the University's mission to contribute to society through the pursuit of
education, learning and research at the highest international levels of excellence.

www.cambridge.org
Information on this title: www.cambridge.org/9781009232241

DOI: 10.1017/9781009232258

First published 2024

Printed in the United Kingdom by CPI Group Ltd, Croydon CR0 4YY

A catalogue record for this publication is available from the British Library

A Cataloging-in-Publication data record for this book is available from the Library of
Congress

ISBN 978-1-009-23224-1 Paperback

"Yu's 'close reading' of life science visualizations goes beyond 'understanding' by shining a critical lens on common forms: graphs, photographs, animations, video games, etc. She goes further than applying Tufte's rules for *Beautiful Evidence* by dissecting authors' intentions to purposely conceal, divert, deceive, misconstrue, and misrepresent, and she embraces practices of open science, transparency, and aesthetics that will help readers experience 'the triumph of interpretation' and empower readers to construct the meaning of heterogeneous, complex data. Her vision is an important counterpoint to the mindless proliferation of chartjunk produced by misuse of common software and moves us to embrace communication practices that not only make for better science, but also serve to enhance our appreciation for the importance of maintaining a healthy skepticism of potentially spurious claims."

John R. Jungck, Center for Bioinformatics and Computational Biology and DENIN Delaware Environmental Institute, USA

"We humans have been using visuals – maps, charts, graphs, photographs, and other types – to see through the complexity of the world and help us reason better about it for ages. However, we tend to take visuals for granted, or to think that they are 'objective' representations of reality, showing it as it truly is. *Understanding Visuals in the Life Sciences* will disabuse you of those ideas. Visuals are intentional, rhetorical arguments and, as such, their quality may be affected by their creators' knowledge, skill, or biases. This book shows how to acknowledge and embrace this notion; doing so may help us become better and more ethical designers."

Alberto Cairo, Knight Chair in Infographics and Data Visualization, University of Miami, USA; author of *The Art of Insight*

"Dr. Han Yu crystallizes the current state of knowledge about visual communication in science through her new book, *Understanding Visuals in the Life Sciences*. Illustrated with key examples both historic and contemporary, she debunks stubborn myths about how publics interact with symbolic rhetoric, while gently urging experts to consider best practices when designing, editing, and producing all genres of visual science communication. This book is a wonderful addition to Dr. Yu's scholarly contributions to multiple fields, expanding the intended audience to the public through her engaging, straightforward style."

Kathryn Northcut, Missouri University of Science and Technology, USA

"Han Yu's *Understanding Visuals in the Life Sciences* is comprehensive in its scope of visual communication, richly illustrated, and amazingly accessible and easy to read. Her insights into visualizing science are cogent, clear, and convincing, and they will greatly benefit scientists and their public audiences, as well as communication scholars and their students."

Charles Kostelnick, Iowa State University, USA

# Contents

# Foreword

"A picture is worth a thousand words" is an expression used to state that a single image or visual can convey a complex idea more efficiently than a verbal description. Isn't it more efficient to show a picture of your new car, rather than to describe it to your friends? Isn't it easier for the police to identify and find a suspect from a photo, rather than from the descriptions of the witnesses? Indeed, as often shown in crime films, the police involve sketch artists to transform witnesses' descriptions to a drawing. And so on. The assumption behind all this is that pictures, photos, drawings, and any other kinds of images "speak for themselves." They are supposed to represent reality not only much more efficiently than words, but also as things really are. So, you might think, reading this book is about something that we already know? That visuals in science are helpful because they help us "see" what is otherwise impossible to observe? Well, not really. In fact, as Han Yu shows in this brilliant and engaging book, no kind of visual speaks for itself. Visuals are extremely useful for communication both among scientists and between scientists and the public, broadly conceived, but only under certain conditions. The recipients of the message need to know how to read it, and the senders of the message need to form it appropriately. Perhaps counterintuitively, Yu argues, no kind of visual is "objective" or "self-explanatory." In contrast, subjectivity is inherent in all the ways that visuals are created and interpreted. A picture can indeed be worth a thousand words, but only when both the sender and the recipient are on the same page. On whichever side you find yourself, reading this book will make you create or perceive visuals in the life sciences in an entirely new way. You will realize this the next time you come across a visual, or start creating one.

**Kostas Kampourakis, Series Editor**

# Preface

As other books in the *Understanding Life* series, the present book highlights misconceptions, preconceptions, misunderstandings, and stereotypes that some of us bring to understanding life science visuals.

*Unlike* other books in the series, the present book probably needs to first address the question of "why bother?" Other books in the series tackle topics that instantly appear complicated and, for the lack of a better word, "scientific": genes, DNA, intelligence, cancer. By virtue of being highly scientific, it makes sense that they would invite misconceptions or misunderstandings and thus warrant a book. "Visuals," by contrast, seem the easy part. I mean, they are simply the things that we see, aren't they? Those of us who are not formally trained in life sciences may not understand scientific jargon or complicated theories, but surely we can *see* a picture and get that.

Well, that notion, my friend, is itself a misconception. Or, more precisely, it represents a series of related misconceptions about science visuals: that visuals are easy, that they represent the reality, that they require no special training (or less training) to understand, and that all it takes is for us to open our eyes and see.

It is understandable that we should see scientific visuals this way, as something easy, or easier than scientific writing. Language, especially written language, requires diligent learning to master. By contrast, visuals seem something that children can naturally pick up and start creating without purposeful instruction. And before they can read "real" books, children as young as two years of age can enjoy picture books.

Similarly, in college education, which is the context I work in, we often require science students to take one and sometimes multiple writing classes to hone their language skills. In science writing classes, students learn to define scientific jargon, articulate scientific theories, and present scientific evidence. In all of my years in higher education, I have yet to teach (or even know of) an undergraduate class that focuses specifically on science visuals. The only way to teach visuals is to teach them as an "add on" in writing classes, as sort of a visual section. This educational structure, then, fosters the perception that visuals come naturally to scientists, that they are easy.

Our general life experiences help to foster this perception. It goes without saying that ours is a visually dominated world. From 4K videos to animated cinema, vibrant color displays to stunning special effects, we are used to seeing spectacular, complex visuals. By pure familiarity and exposure, we have come to think that we understand or should understand them.

It is the argument of this book that all of the above are misconceptions. Life science visuals are far from being simple or easy. They are complex rhetorical devices – as complex as language. They are not easy to create – if we want them done right. They are not easy to understand – especially for a nonspecialist audience. Throughout this book, I use specific life science visuals to demonstrate these points.

Hopefully, by doing so, by broaching the misunderstandings and misconceptions surrounding life science visuals, the book can inspire visual creators – be it scientists or science communicators – to be more conscious in creating useful visuals for the public. At the same time, a nonspecialist audience may gain a deeper and more sophisticated appreciation of life science visuals, may feel more empowered as they approach these visuals, and may be more willing to engage with and question these visuals.

But first, I need to narrow down the scope of this book. When we speak of science communication, that "communication" is not just one type of communication. Depending on the parties involved and the purpose of the communication, there are multiple types of science communication.

There is communication between fellow scientists and specialists, such as one biologist writing a research paper published in an academic journal to be perused by fellow biologists working on related research topics. Think of this as professional science communication. There is also communication between trained scientists and scientists in training, such as a biologist writing a biology textbook for college students. Think of this as classroom science communication. Then there is communication between scientists/science writers and the general public, such as Physiology/Medicine Nobel Prize winner Stanley Prusiner writing a book about the discovery of prion diseases. Think of this as popular science communication.

Given the public-facing nature of the *Understanding Life* series, the focus of this book will be visuals used in popular science communication. When relevant, the other kinds of science communication and visuals used in those contexts will be discussed, with the primary goal of illuminating visuals used in popular life science communication.

# 1 Introduction

For a book that attempts to explain how to understand visuals in life sciences, it seems prudent to first explain what we mean by "visual," even if it may seem quite a common word.

In everyday conversation, "visual" is often used as an adjective and means "relating to seeing or sight," as in "visual impression" or "visual effect." In the context of this book, "visual" is used similarly as an adjective, but in addition, and more often, it is used as a noun. As a noun, it refers to the variety of images used in life science communication. For example, photographs are a type of visual commonly used in life science communication, and so are drawings.

I understand that, as a noun, the word "visual" is decidedly more awkward than, say, "image" or "picture." However, people who study visual representations like the word because it is a more generic concept, an umbrella term. Other words tend to be narrower in scope or have specific implications. For example, "picture" too strongly suggests a photograph and a photograph alone, whereas "image" may remind one of something that is entirely imaginary, as in a "mental image."

## The Importance of Visuals in Life Sciences: A Brief History

Visuals are essential in life sciences – not only for communicating scientific findings and discoveries, but also for carrying out scientific research in the first place. They are part and parcel, you may say, of the science.

Think about it: Scientists rely on imaging technologies to observe and make sense of the research subjects they study. Looking at a virus under the microscope, for example, allows scientists to understand the virus's structure and, from there, to explore its function. Photographing a plant from flowers to leaves allows scientists to make accurate identification of the species and, from there, to assess its cultivation. Without being able to visualize the life around us, there can be no life sciences.

When scientists or science writers are ready to communicate research findings, visuals take on multiple roles. They appear in print publications, on TV, on the internet, in the classroom. Some of these visuals serve as scientific *evidence*: If one has observed a certain unique viral structure under the microscope, what's better evidence than a detailed micrograph to show the said structure? Some visuals also serve as *argument*: If one has a hypothesis for how a virus spreads, putting that process in a simplified diagram, as opposed to lengthy texts, can help people grasp the gist of the argument and, potentially, be persuaded of its soundness.

The importance of visuals in life sciences is not a recent phenomenon either. The rise of life sciences itself, one might say, is connected with the rising popularity of visuals. A well-known testimony to this connection is *Micrographia*, a book published in 1665 by English scientist and architect Robert Hooke (1635–1703). The complete title of the book is a bit long and awkward. It goes *Micrographia: or Some Physiological Descriptions of Minute Bodies Made by Magnifying Glasses. With Observations and Inquiries Thereupon*.

In the book, Hooke presented the minute bodies he saw using the microscopic technologies available to him at the time. With 38 stunning copperplate engravings, *Micrographia* illustrated the structure of crystals and plants. But what made the book eternally famous are, decidedly, its bugs: ants, fleas, flies, lice, you name it. The fascinating – or repulsive, depending on whom you ask – details of these bugs are lavishly rendered in the book. The depiction of a flea, for example, unfolds to more than twice the size of the book (Figure 1.1).

A century earlier than giant fleas, similarly stunning depictions of the human body had been rendered. In 1543, Andreas Vesalius (1514–1564), a Belgian

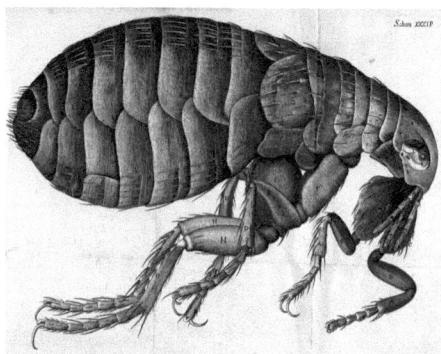

**Figure 1.1** Robert Hooke's depiction of a flea.

anatomist and physician, presented to the world *De Humani Corporis Fabrica Libri Septem (Seven Books on the Fabric of the Human Body)*. The volume was considered one of the first, if not *the* first, accurate depictions of the human body.

Anatomic accuracy is only part of the volume's fame. As *Micrographia*, *De Humani Corporis Fabrica Libri Septem* is a spectacular visual feast. Included in it are elaborate woodcut illustrations of human dissections. The human bodies don't simply lie lifeless on a dissection table. Rather, they are stood in dramatic poses against a landscape, as if erected pieces of classical statues. They are known as the "muscle men." The men's bodily tissues are stripped away layer by layer in each passing plate to reveal the anatomy of the human body (Figure 1.2). Meanwhile, the landscapes of the individual plates form a continuing panorama. As the muscle man is stripped clean and eventually collapses, the life behind him goes on.

If you think sixteenth- and seventeenth-century scientists are a bit sensational with their giant insects and theatrical humans, today's scientists outdo them with pure technological prowess. Modern life science research frequently deals with things that are not only invisible to the naked eye, but also difficult to conceptualize in the first place: DNA, genes, genomes. To visualize such concepts, scientists rely on sophisticated technologies, from X-ray crystallography that

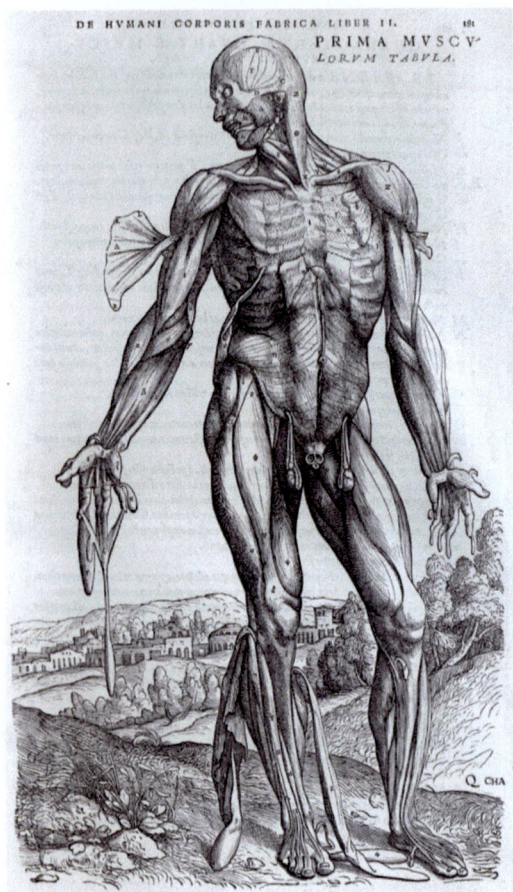

**Figure 1.2**  Andreas Vesalius' depiction of the muscle men.

reveals DNA's structure, to heat maps that show levels of gene expression through color-coded boxes (Figure 1.3), to sequencing technologies that turn genomes into endless scrolls of letters.

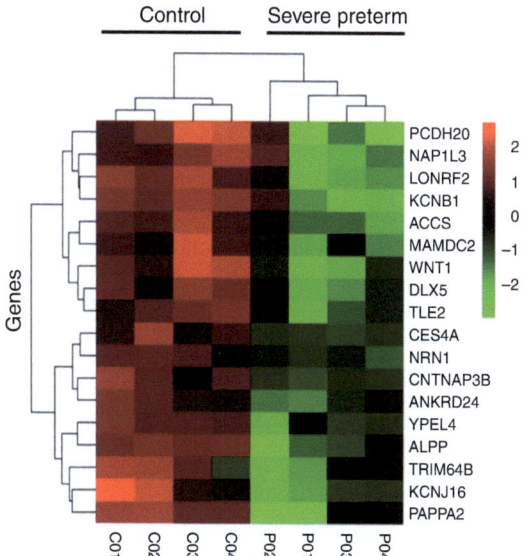

**Figure 1.3** A heat map.

In Figure 1.3, each row represents a different gene, and each column represents a different test sample. The color of each little box stands for the change in gene expression for each gene in each sample. Red shows increased, or upregulated, expression – the brighter the red, the higher the expression. Green shows decreased, or downregulated, expression – the brighter the green, the lower the expression. Based on the gene expression patterns, we can then attempt to group similar genes and samples for study.

## The Changing Nature of Life Science Visuals

As the above brief history shows, visuals have long been and will continue to be paramount to life science research. What is not yet apparent is that different visual "styles," if you will, have come and gone. Yes, the advance of scientific methods and technologies necessarily results in different kinds of visuals. But that's not all. The nature of the visuals, the value we attach to them, as Galison explained, have also evolved.

Before the nineteenth century, natural sciences saw the function of their visuals as one of revealing "truth to nature." This "truth," in case one wonders, is not the contemporary notion of "objectivity" that is often associated with scientific efforts. In fact, in some ways, it is the opposite of the modern concept of objectivity. "Truth to nature" believes that we can't grasp the true nature by observing and recording individual plants or animals as they are, because individual specimens often contain idiosyncrasies and imperfections. The leaves of a plant may have been chewed on by an insect; the wings of the insect may have been notched by a predator. Recording these realities, it was believed, doesn't reveal truth.

To get at the truth of nature, we must rise above individual specimens and their idiosyncrasies. How do we do that, you ask. By relying on geniuses – geniuses such as Robert Hooke and Andreas Vesalius. Their artistic and scientific talents allowed them to see past, through, and beyond individual specimens and to create perfect, awe-inspiring fleas and men. This is why their visual creations have that unmistakable touch of artistic invention, even divine intervention. They say, as it were, *Look at this specimen, so symmetrical, so wonderful, so ideal. Make no mistake. They are not accidents. They are God's creations.*

Around 1830, the "truth to nature" ideal, though not extinguished, started to give way to the concept of "mechanical objectivity." The difference between the two concepts was stark. While "truth to nature" relied on artistic correction and invention to render truth, "mechanical objectivity" sees artistic correction and invention as contaminations of scientific data. The role of visual creators, the objectivity approach announced, is not to idealize or editorialize what they see, but to simply record what they see, warts and all. And if that means drawing a plant with gnawed leaves, so be it. Only such work can let nature "speak for itself."

Under this belief, scientists were expected to practice a strict moral culture. They must not let their imagination run wild, they must not invent details to fit their theories, and there will be absolutely no retouching of any visuals, even if it is to sharpen them to make them more legible. Scientists were to be copy machines through which nature reveals itself. In fact, if possible, using real machines would be even better, because they give us the ultimate mechanical objectivity. Real machines, such as a camera, are not motivated by personal interests, are free from human interventions, and can therefore create the most objective visuals. These beliefs, as we will see later, are flawed. But that's for the rest of the book.

Although the pursuit of objective visuals is still a relevant goal in today's life sciences, it is no longer the only or the most important one. The newest kid on the block, starting to make its appearance in the early twentieth century, is interpreted visuals. As their name suggests, these visuals are a result of interpretation, not mechanical copy-making. Whose interpretation, you ask? Experts, trained scientists. The idea is that these experts have accumulated a vast amount of experience and knowledge; they have observed hundreds, if not thousands, of the same specimens. While each individual plant of the same species may look a bit different – a little bigger or smaller, a little crooked or bent – the experts can synthesize the differences and create something that captures the essence of the plant.

Compared to mechanically objective visuals, interpreted visuals have some distinct advantages. Imagine medical students who are being trained to recognize brain lesions using radiographs of real patients. Given a group of inexperienced students, the chances are that the lesions will not jump out at them. In other words, the task of looking *at* something and learning to recognize it will become a task of looking *for* something, not knowing what one is looking for. The "objective" radiograph, then, would lose its instructive value. By contrast, an interpreted drawing of the radiograph can tone down irrelevant details and exaggerate the essential features of the lesions. This way, the visual can instill in students the basic knowledge to identify similar, but less obvious, lesions in actual patients.

## Challenges with Using Visuals in Popular Science Communication

Before the rise of modern science circa seventeenth century, there was not a clear demarcation between science communication meant for professional scientists and communication meant for the wealthy, the intellectual, and the generally curious. Robert Hooke's *Micrographia*, for example, was a bestseller and inspired wide public interest.

One can therefore say that, as the above history shows, visuals have been used in popular science communication for a long time. That tradition has only intensified in contemporary popular life science communication, as we will see throughout this book. The advent of digital technologies and online publishing assisted with this trend: Engraving or printing visuals in a book are costly, but in a blog or ebook the added cost of visuals is minimal.

But does this mean that all these visuals are actually helpful to everyday readers? Is each picture, as we are wont to think, really worth a thousand words?

For specialist audiences such as trained microbiologists or neurobiologist, visuals within their specialties often make automatic sense – because these visuals are an ingrained part of their training and have become, so to speak, common knowledge. The mere appearance of a visual can potentially reveal how it came about, why it was made, what technologies were employed to make it, what the colors denote, what the scales suggest, etc.

For nonspecialist readers, that's often not the case. As Trumbo wrote about DNA and the seemingly well-known image of the double helix,

> Most of us have seen images of the double helix, that wonderful spiral strand of "something," first proposed by Watson and Crick. What is its scale? Is DNA larger than a drop of water, a molecule, an atom? What is its color? Does it really look like a spiral? This animated, spiral-shaped image fills the television screen during a report about cloning. It spins clockwise, or is it counterclockwise? It appears in advertising for a biotechnology company wrapped around a typographic logo. It appears in a lively cartoon sequence within the movie *Jurassic Park*. What exactly are we seeing? (p. 381)

As this quote implies, the double-helix visual plays multiple roles in the public sphere: It is used to explain genetic research; it is turned into cinematic entertainment; it is an endorsement of commercialized biotechnology. Can the same visual fulfill these very different roles and purposes? What information does the public gain when they look at the visual? What visceral reactions do they form? If readers cannot determine the scale of DNA from a spinning double helix, is the visual really helping to illustrate what DNA is and does? Should we add more technical details to the spiral to make it more informative? Is that going to help? Or is it going to make things worse? And what do we mean by "help" or "make things worse"?

These are complex questions that don't have easy answers. But for that reason, they are precisely the kinds of questions that are bound up with the misconceptions and misunderstandings surrounding life science visuals. And they are the kinds of questions this book ponders.

## But Hasn't This Work Been Done?

Surprisingly, although a lot of work has been done on life science visuals, very little of the kind that I proposed above has. Existing studies often focus on the visual practices and needs of professional scientists. For example, they look at how scientists create persuasive visual evidence in their publications. Or, they discuss what visual technologies scientists need in order to better present the increasingly complex and large data sets they obtain.

In addition, existing studies also look at the experiences and needs of science educators and science students. Here, we learn that visuals used in science textbooks leave quite a bit to be desired. They have poor data mapping, contain inadequate captions, and use surface features that may mislead students. To their credit, these studies recognize that reading science visuals is a complex process that must be taught to students for students to be truly integrated into the world of science.

All of these studies are of course important and useful in their own ways, but they have largely overlooked you and me, the everyday readers. The general public audiences are not professional scientists who create visuals for publication, nor do they partake in formal education to learn under the guidance of an instructor. Rather, as a National Science Board survey indicates, most American adults gain information about science through informal channels such as the internet, television, newspapers, and magazines.

Given the importance of visuals to the production and communication of life sciences, it is high time that we scrutinize their use in popular communication.

In fact, I would argue that, to obtain wider impact, it is more important that we pay attention to the vast number of public readers out there than to the limited number of professional scientists and scientists in training.

This is true for all disciplines of science, but especially true for life sciences, which are intimately related to people's everyday lives and personal well-being. Take the discipline of genetics as an example. Classical genetics from the early twentieth century offered a way for farmers to better control their crops and boosted the agricultural industry. The infamous eugenics movement in the 1930s, in trying to "regulate" human birth and development, condoned involuntary sterilization, often forced on minority populations. Today's genetics research holds a lot of promise for human health and welfare, from genetic testing to genetic therapy, though it is also at the center of much controversy, from genetically modified food to designer babies. Beyond genetics, other life science research also has profound social implications, from the preservation of diverse species on Earth to the elimination of fast-spreading, deadly viruses.

With so much at stake, the public needs to be actively involved in understanding and engaging with life sciences, and scientists need an informed and engaged public that supports sound, ethical scientific efforts.

## The Deficient Public

Now that I've mentioned the importance of public understanding of and engagement with science, I must add that "public," "understanding," and "engagement" are all loaded words in the context of popular science communication research.

To start, the word "public" tends to convey the impression of the public as a homogenous, united front. But this is certainly not true. In reality, the public consists of diverse communities, groups, and individuals with varying education, interest in science, trust in science, political stances, life values, etc. It is beyond the scope of this book to account for all these variations, but I do want to acknowledge the fact that the public is a plural concept and invite all readers to do the same.

Next, the words "understanding" and "engagement" are loaded because they carry a lot of historical baggage when we consider the relationship between the public and science. The traditional view of this relationship, to put it bluntly, assumes that science (and by extension, trained scientists) are superior, and that the general public has a lot of catching up to do. Scientists are the ones (and only ones) who have the ability and duty to educate the masses and teach them essential scientific concepts so that the public can become literate enough to comprehend and appreciate the gist of modern science. Once sufficiently literate, the masses stand a chance to be rational and productive citizens in the modern society – and they will trust science and support scientific research.

Studies that are grounded in this approach often use standard surveys to reveal the public's poor grasp of basic scientific concepts (such as their inability to define what counts as a "scientific experiment"). Based on these survey findings, the studies then recommend proper education. For example, people are advised to take at least some college-level science classes.

This perspective, the so-called "deficit view" – the public being the deficient party – dominated in the 1970s, 1980s, and into the 1990s. In fact, it still holds sway today among some scientists. But we have now come to recognize that this view is problematic. It is based on a very rigid, elitist view of what counts as "expertise" and thus a very rigid, elitist view of who has expertise and who doesn't. Everyday readers may not be able to technically define what counts as a formal scientific experiment, but that doesn't mean they can't recognize it when they see one, or that they all lack the ability to think logically, to weigh evidence, and to make sense of scientific findings.

Brian Wynne's work on Cumbrian hill sheep farming is one of the most well-known studies that demonstrate these points. In 1986, after the Chernobyl Nuclear Power Plant exploded, thunderstorms washed nuclear fallout onto Britain's upland Cumbrian area. The fallout contaminated the soil and vegetation and threatened to contaminate the sheep that grazed in the area. In the aftermath, scientists from the Ministry of Agriculture, Fisheries, and Food tried to assess and reduce sheep contamination in the area. With their rigid understanding of what a perfect scientific experiment should look like, these experts ignored farmers' knowledge of the local environment, how the hill sheep behave, and the reality of hill farming management. As a result, much of the scientists' formal assessment and experimentation failed.

For example, at one point, the scientists wanted to conduct an experiment on the ability of the mineral bentonite to absorb contaminants from the land and reduce sheep contamination. To do that, they marked out different plots, spread different concentrations of bentonite in each plot, and left some plots untreated as controls. Different sheep were then assigned to graze in different plots and periodically tested to assess the effect of the mineral. The farmers pointed out to the scientists that the experiment as designed wouldn't work. Hill sheep are used to roaming open lands. If they are fenced in, they would waste away, which would ruin the experiment. The scientists ignored the farmers' concern, deeming it irrelevant to the rigor of scientific experiment. But soon enough they quietly abandoned the experiment for the precise reason the farmers had identified.

## Engaging the Public

Studies such as Wynne's gave rise to a new approach to map the relationship between the public and science, one that may be called the engagement approach. This new approach emphasizes "dialog" and "engagement" rather than "understanding" – because the word "understanding" always already implies deficiency on the public's part. The new approach asserts that science is a social enterprise and that everyday people's life experiences, local knowledge, and values matter. Simply providing the public with more scientific facts isn't going to turn them into believers or supporters of science. The public is not a problem to be fixed, but a party to be actively involved and engaged in science.

While the engagement approach has come a long way from the deficit approach, it is not without problems of its own. First, there is no specific agreement on what exactly counts as "engagement," or how it can be made effective. Some suggest trying to stimulate the public's curiosity and sense of wonder about science; others encourage shared decision-making between citizens and scientists; still others focus on facilitating direct conversations between the public and scientists. Without effective guidelines, engagement efforts risk becoming mere tokens.

Second, talks of public engagement risk creating a romantic view of the public, an ultra-version of the sheep farmers in Wynne's study, if you will: inherently wise, possessing knowledge uniquely relevant to scientific research, and having pragmatic values that are superior to the narrow agenda of trained scientists. Pushed too far, this romantic view can render scientific training irrelevant. Also, the public is made out to be a homogenous group when, in reality, not all members of the public will have the same amount of interest, knowledge, or experience about science or the same amount of trust in science.

Finally, the engagement approach continues to create, if only inadvertently, a split between science and humanities, scientists and citizens, hard facts and soft feelings. The public remains the public, while the science is, well, the science, and one is not like the other.

## Where Do We Go from Here?

While it is relatively easy to critique existing approaches, it is much harder to figure out a perfect alternative. The relationship between science and the public is complex! With that in mind, it is a start to recognize that we ought to consciously bridge rather than separate the two, to learn to think of them as the inherently connected entities that they are, or should be. After all, why develop science if not to serve the public?

The next step is to recognize that neither science nor the public is a singular concept. Scientists come from different disciplines, have varying training experiences, and enjoy diverse cultural backgrounds. They don't think the same way about the role of the public or the value of popular science. Many of those who communicate science to the public are not trained scientists either. They are science writers and journalists who may or may not have a formal science background. Similarly, the public is an inherently variable group; or, rather, each individual is an inherently multiple identity. We are patients, consumers, taxpayers, parents, etc. Not all of us think or feel the same way about science or specific science topics, and the different hats each of us wears tug us in different directions.

Given all these variations, what is likely to work, as Holliman et al. argue, is not a "one size fits all" approach to popular science communication, but a fluid, context-specific approach. In some communication contexts, we may want to empower the public, acknowledge their local knowledge, and encourage two-way communication or public debate. In other contexts, we may need one-way delivery – via TV, newspaper, the internet – of the latest and newest scientific findings, fashioned in ways that meet the needs of a public audience. And in many contexts, we may need a combination of both, mixed at different ratios.

What this means for science communication research is to forget about lofty theories that try to generalize broadly, but instead examine specific cases in detail to glean nuggets of useful information.

The same, as I attempt to show in this book, applies to studies of popular life science visuals. It is through specific examples that we can appreciate the importance and challenges of using visuals in popular science communication. It is through specific examples that we can hope to dispel misconceptions and stereotypes and gain lessons learned.

More fundamentally, it is through specific visual examples that we can try to bridge science and society, to see that life science visuals are more than some data that came out of a science lab, that they are also products of social and cultural contexts, or they should be. From there, we can start to appreciate the multiple functions and effects these visuals have.

## What Kinds of Visuals Will We Examine?

In the rest of the book, I apply all the above to practice. The misconceptions and stereotypes about life science visuals are discussed against the backdrop of the changing nature of science visuals and the different ways to think about the relationship between science and the public. To do so, I have selected the seven most commonly used kinds of visuals in popular life science communication. Each type is described and discussed in one chapter.

The first type of visual is photographs, the topic of Chapter 2. Photographs are commonly used in life science communication. Too often, they are seen as objective records of reality, as machine-captured visual evidence where the pixels *are* what they are. Chapter 2 confronts this perception. It concedes that photographs *can* serve as observational records, but they are not free from human manipulation. Machines, after all, are set and operated by humans, and anything from exposure time to shutter speed can affect the outcome. There is also the purposeful selection of what scenes to capture, what scenes to ignore, and post-editing enabled by software such as Photoshop.

Moreover, in today's popular science communication, photographs are often purposefully used for their ability to evoke strong viewer reactions and emotions. Examples of such photographs are abundant in the media. For example, photographs of Frankenstein food – such as vegetables and fruits taking a shot of colorful fluid straight from a syringe – are designed to incite public outcry against genetically modified food. These and other such photographs have their value in conveying social and political attitudes, but that is precisely it: This intention and effect need to be known so that public viewers do not take these symbolic photographs as scientific evidence.

The second type of visual I discuss is micrographs. Chapter 3 provides a concise history of the use of micrographs in life science communication, from hand-drawn microscopic observations to scanning electron micrography. This history demonstrates our obsession with seeing ever more minutely, more clearly, and more reductively – that is, we seem to believe that when we break nature down into its smallest parts, we *will* understand it. But, in reality, a reductive focus can pose a challenge to nonspecialist viewers because microorganisms such as the SARS-CoV-2 virus have no counterpart in the everyday visual world we live in. Without contextual cues, then, nonspecialist viewers gain very little by looking at a static, ultra-zoomed-in micrograph. Depending on the microscopic techniques used, a virus can also take on diverse appearances, which may cause more confusion than clarification.

In addition, as with photographs, micrographs are too often considered a realistic portrait of microcosmic life. But, in reality, creators of micrographs enjoy quite a bit of creative license. For example, colors are used exuberantly, not entirely or even necessarily to enhance meaning, but to create emotional reactions. This is particularly ironic when we consider that viruses and most cells do not have color and, in that sense, are free of the emotional baggage we grant them. Yes, artistic micrographs serve to celebrate life as the beautiful and mesmerizing thing that it is. At the same time, they are bound up with science's need to promote itself, to portray itself as a similarly beautiful and mesmerizing enterprise.

The third type of visual, and the focus of Chapter 4, is illustrations, or drawings, a staple in life science communication. In the year 2020, the single most popular visual in the media is the gray–red, fuzzy-looking SARS-CoV-2 virus developed by the US Centers for Disease Control and Prevention. Despite being commonplace, scientific illustrations are in many ways a blackbox: They mask the creative – and scientific – decisions that go into making the illustrations and present an end product that says, as it were, this is the way life *is*. The use of precise lines, explicit shapes, and other ultrarealistic details all help to convey this visual certainty.

It is, then, easy to forget that illustrations are interpretive visuals, or someone's interpretation of life. For example, a colorful, textured, three-dimensional image of a protein is not reality; it is an artistic rendering based on deduced atomic structures. While these artistic choices can create visual appeals and enhance understanding, they also risk becoming "seductive details" that distract readers and create false impressions. A case in point is visual metaphors – illustrations that reduce sophisticated concepts into smart-looking yet ambiguous metaphors, such as drawing a tree and morphing its branches into the DNA double helix. Not only are these images technically inaccurate, but their metaphorical meanings are vague. At worst, they promote false security in viewers that science is simple, familiar, and *not* worth asking specific questions.

The fourth type of visual discussed in this book is graphs, which are commonly used in life science communication to convey quantitative data. Chapter 5 starts by overviewing the most common types of graphs, such as bars, lines, pies, and histograms. It discusses these graphs' common usage in life science communication. The chapter also discusses pictographs, including their history, growing popularity, and advantages, as well as the challenges of using them in life science communication.

The main argument of Chapter 5 is that graphs, contrary to what some visual creators may believe, are often difficult for nonspecialist readers to understand. For many viewers, the scientific concepts and quantitative data being graphed are abstract or unfamiliar to begin with. Certain graphic conventions and stylistic choices make the matter worse, such as the use of three dimensions, truncated axes, multiple *y*-axes, intrusive details, and a lack of integrated verbal explanation. Given these barriers, graphs that are meant to synthesize a large amount of data may end up confusing rather than enlightening viewers. In fact, these graphs can leave public readers with a heightened conviction that science *is* an inaccessible enterprise. Worse still, graphs that are purposefully designed to encourage a certain reading may mislead viewers.

The fifth type of visual, and the focus of Chapter 6, is interactive visuals, which are web-based visuals that allow direct user interactions such as clicking and dragging. Life sciences are big data sciences, and when it comes to reporting big data, online, interactive visuals have inherent advantages. They can present enormous amounts of data for correlation and comparison. They do not overwhelm viewers by presenting everything at once. Depending on user input, select data can be visualized on demand.

At the same time, interactive computer technology alone does not equate to effective visual communication. Not only do interactive visuals need to be usable – that is, pressing a button is easy and will result in a certain action – they also need to be useful. That is, they need to support the kind of action and output that users desire. If users want to click on one specific city in an interactive map to find local epidemiological data, a map that is chock-full of data but prevents targeted selection is, strictly speaking, useless.

Games offer another type of interactive visuals. Well-known examples in life sciences include Foldit, a protein structure prediction game, and Phylo, a multiple sequence alignment game. These games take advantage of humans' superb visual recognition ability to solve computationally expensive tasks such as arranging DNA sequences, and they have been hailed as great successes of public engagement in science. However, on closer inspection, these games often do not engage players in anything scientific *per se*, beyond using their labor. Whether or not that is okay is an interesting question to ponder.

Last, in Chapter 7, we turn our attention to infographics. Infographics have gained tremendous popularity in contemporary popular science communication. They are used to illustrate anything from global population change to the biomechanics of a running cheetah. But infographics are quite a nebulous concept. They are essentially information in graphic formats. As such, infographics can be as simple as a pie graph with some text positioned around it, or as complex as a constellation of graphs, illustrations, photographs, and text.

If there is one overarching difference between infographics and other types of visuals, it is that infographics, with their multiple and combined use of visuals and text, are more likely to enhance understanding yet at the same time just as liable to cause confusion. The various misconceptions concerning other types of visuals have compound relevance in infographics. Because of this, visual creators need to be extra conscious – and conscientious – of their designs, and viewers need to be extra critical of their interpretations.

In short, we have quite a spectrum of visuals to cover – and quite a spectrum of life and science to go along with them. I hope you find the journey ahead an intriguing and informative one.

# 2 Photographs

Photographs are often considered an "easy" and accessible type of scientific visual. After all, they are commonplace in everyday life and not exclusive to scientific research. Everyone takes photographs and knows what photographs are. As long as one can physically see, one (so it is thought) can get what a photograph is about. Unfortunately, when it comes to life science photographs, much of this is misconception. This chapter explains why.

## A Brief History of Photography

*Camera* is a common word in today's English language. We all know what device it refers to and what that device does. My phone, for example, has a camera; in fact, it has several cameras with different lenses to take photos with different effects. But did you know that the word *camera* originally meant "chamber," as in *camera obscura*, which is Latin for "dark room" or "dark chamber"?

The camera obscura is the old, old ancestor of the present-day camera. The device consists of a darkened room or enclosed box with a small hole on one side. When light passes through this hole into the room or box, it projects a picture of the outside world onto the opposite wall or screen. The earliest record of the camera obscura effect dates back thousands of years to ancient China (circa 400 BC). In the early sixteenth century, Leonardo da Vinci (1452–1519) provided the first clear description of the device, including sketches, in his book *Codex Atlanticus*.

A camera obscura allows a viewer to see the projected image in real time, but it cannot preserve that image for later viewing or take the image away from the scene to other viewers. Three hundred years would pass after da Vinci's sketch before inventors found that chemicals such as silver salts and bitumen of Judea (a naturally occurring asphalt) can create a lasting record. When exposed to light, these chemicals darken or harden, thereby fixing the projected image onto a plate. The world's first photograph was created this way in the early nineteenth century by French inventor Joseph Nicéphore Niépce (1765–1833) using a camera obscura, a bitumen of Judea coated plate, and 8 hours of exposure.

Later inventions such as iodized silver plates and collodion-coated glass plates sped up exposure time, allowed repeated copying from a single negative, and enabled mass-production of prepackaged plates and then rolled film. With these advances, photography became accessible to everyone, not just people who had a working knowledge of chemistry.

Of course, even with all these developments, photography wouldn't have been the success it was if not for colors. The early photographs were simple black, white, and gray, and people got over the excitement pretty quickly. If only the vibrant colors in nature, in their households, and in their clothing could be captured in photographs. That wish took a long while to come true.

The first practical and commercially successful color photographic technique – known as the autochrome – was invented in the early twentieth century. Autochrome works by spreading millions of tiny, invisible grains of potato starch onto a plate. These grains were dyed red, green, and blue, and acted as color filters before the light hit the light-sensitive material on the plate. Following exposure, the plates can be processed to produce a color photograph. Subsequent inventions led to improvements such as multi-layered color films with improved color sensitivity and reduced cost, making color photography a norm rather than a novelty.

## Early Life Science Photographs: What You See Is (Pretty Much) What You Get

The moment photographic technique became presentable, science fell in love with it. The ability to record and catalog plants, animals, and various life forms as they appear provides an invaluable way to build and spread knowledge.

This method is far better, it was believed, than engraving or woodcuts. As a New York doctor, Ransford van Gieson, exclaimed in 1860 in the *New York Medical Journal*, photography can "secure accurate representations" and "present the exact appearance" of specimens, including "the most rare and curious specimens of disease."

Examples of such early science photographs abound. Around 1843, English botanist and photographer Anna Atkins (1799–1871) published her book *Photographs of British Algae: Cyanotype Impressions* (Figure 2.1). Atkins was widely considered the first person (a woman no less) to illustrate a book with photographic images. The book, as the title suggests, uses the cyanotype

**Figure 2.1**    Anna Atkins' photograph of *Halyseris polypodioides*.

technique. This technique works without a camera. Instead, the specimens are placed directly onto paper or glass coated with light-sensitive materials and exposed to light to create the impression. The resultant photographs (or photograms) have a signature cyan, or dark blue, color.

In the decade that followed, Atkins continued to distribute her work, occasionally adding new plates. With Anne Dixon, Atkins also published *Cyanotypes of British and Foreign Ferns* in 1853 and *Cyanotypes of British and Foreign Flowering Plants and Ferns* in 1854.

Around the same time, French neurologist and physician Guillaume-Benjamin-Amand Duchenne de Boulogne (1806–1875) was busy photographing human faces. Interested in the mechanism of human expressions, Duchenne de Boulogne tried to reproduce various expressions – joy, terror, aggression, sadness, and pain – by electrically stimulating facial muscles. Duchenne de Boulogne photographed his research subjects in their moments of being electrified (Figure 2.2) and published them in *Mécanisme de la Physionomie Humaine* (*The Mechanism of Human Physiognomy*). The book contains 84 photographs and went through two editions (1862, 1876). In some cases, masks were used to cover half of the subject's face so the effect of the electrical stimulation can be compared with the other half. Duchenne de Boulogne eventually concluded that there are specific muscles for the expression of each emotion, and that some of these muscles are more controllable than others.

American scientists and physicians joined in this early adoption of photographs. Their work can be seen in the six-volume *Photographs of Surgical Cases and Specimens*, commissioned by the United States Surgeon General during the Civil War. The book contains about 400 photographic prints with accompanying case descriptions. The photographs depict anatomical specimens such as fractured thighbones and skulls (Figure 2.3), as well as soldiers who had suffered traumatic wounds and subsequent surgeries.

The skull in Figure 2.3 belongs to a confederate soldier who was wounded on July 17, 1864. The soldier was admitted into Lincoln US General Hospital on that day and died two hours later. The skull, as the case narrative explains, shows "multiple depressed fractures of the vault of the cranium . . . The point of

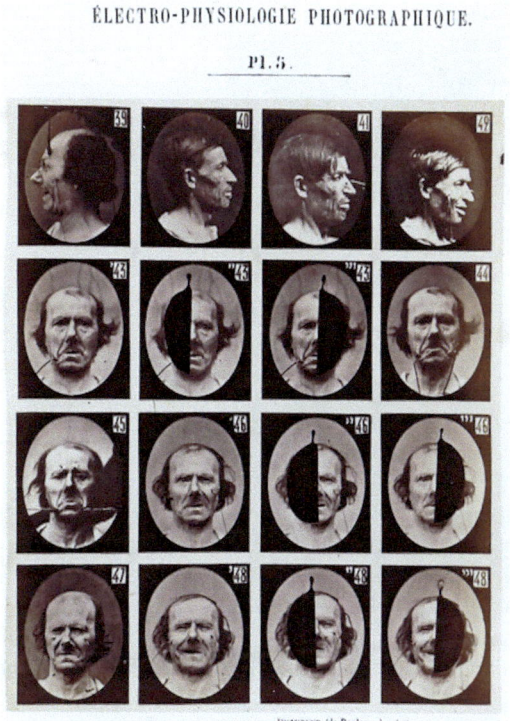

**Figure 2.2**    Plate from Guillaume-Benjamin-Amand Duchenne de Boulogne's *Mécanisme de la Physionomie Humaine*.

greatest depression is an inch to the left of the median line" (volume 1, specimen no. 2871).

These early science photographs may seem unsophisticated compared to today's full-color, high-tech visual splendor, but their value as scientific evidence was remarkable in their own times. As Gross et al. have explained, nineteenth-century life scientists – like earlier natural philosophers – were still doing a lot of observation and fact gathering as opposed to theory-building. Although quantitative measurements – that is, numbers – had started to appear

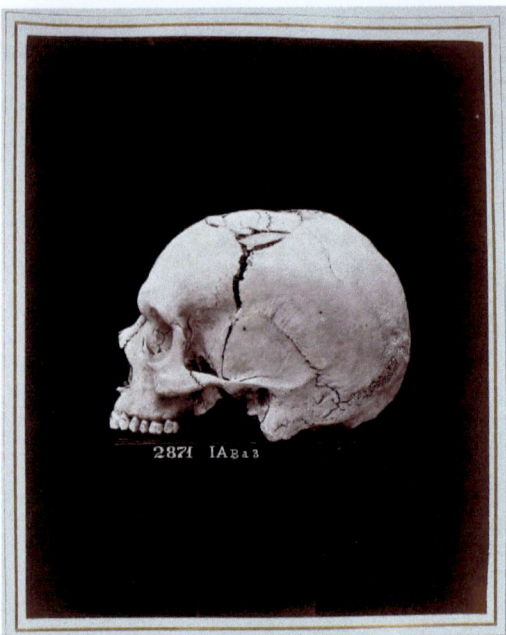

**Figure 2.3**    Cranium extensively fractured by a shell explosion.

to describe said observations and facts, for the most part, qualitative descriptions of color, form, and "feel" continued to dominate.

Because of these realities, photographs had a distinct advantage in the nineteenth-century life sciences. As we can all appreciate, words can only do so much when it comes to describing. Imagine trying to use words alone to describe various algae or human facial expressions. It is an impossible task. There are only so many ways to write about the shape of a blade (the leaf-like structure of algae) or the movement of the mouth before things get utterly confusing. Photographs, by contrast, capture what things look like.

The value of these photographs increases further when we consider that nineteenth-century scientists were preoccupied with mechanically objective visuals, visuals that are free of human manipulation and that allow nature to speak for itself (see Chapter 1). Drawings, engravings, and woodcuts, no matter how carefully and conscientiously done, were considered less objective than photographs, because humans will never be as disinterested as machines.

This, of course, is a naïve view of photography. As we will see later in this chapter, photography is not an automatic mechanical process. For now, though, we have to admit the baseline ability of photographs to capture what something looks like, which is precisely what made them valuable to nineteenth-century scientists. Whether Atkins, Duchenne de Boulogne, or the US Surgeon General, their concern didn't go much beyond systematically cataloging the subjects of their studies and showing their appearances. The photographs are not meant to reveal the *why* or *how* of a specimen. In other words, what the photographs *show* constitutes the value of the scientific work.

And, just the same, readers can immediately *get* the science when they look at the photographs. They can see the details of an alga or a fractured skull. They can also compare different specimens so that they can identify individual ones: one alga has a different kind of blade than another; the stimulation of one facial area creates a different expression than that of another.

Finally, through these photographs, readers get a sense of the scientific method behind the study. Duchenne de Boulogne's photographs show the stimulation of facial muscles, which is the method he used to create different facial expressions. In the other two cases, the method – that of collecting and then photographing the specimens – is implicitly conveyed in the photographs.

In short, with early life science photographs, what you see is pretty much what you get, and what you get is the precise value of these visuals. The troublesome thing is that science changes. In contemporary times, as we will see, the idea that "what you see is what you get" quickly becomes a misconception.

## Contemporary Life Science Photographs: What You See Is (Often) Not What You Get

In contemporary life sciences, the goal of research has shifted to unveiling the why and how of lives around us: how algae can be used to produce biofuel or how a patient with a fractured skull may be saved, as opposed to merely

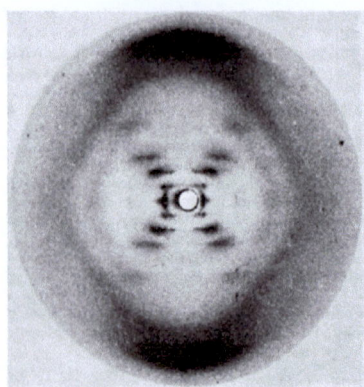

**Figure 2.4**   Photo 51.

recording the existence of the specimens. In addition, contemporary research objects are increasingly small and invisible. We study the cells in the muscles and the myofibrils in those cells, not the mere location of the muscles in the human face. When we make observations, they are increasingly quantitative. We count, measure, and time, not merely relying on how something appears.

As a result, the value of photographs as scientific records and evidence has diminished. Merely "looking at something" is not enough to convey, or to *get*, the science behind it. Photo 51 (Figure 2.4), some say the most important photo ever taken, is a case in point.

Photo 51 was taken by English chemist Rosalind Franklin (1920–1958) and her then graduate student Raymond Gosling (1926–2015) at King's College London in May 1952. The photo, to put it simply, is of DNA. It was not made using an everyday camera; rather, it was taken via X-ray crystallography, which is a method to determine the arrangement of atoms in crystals. Crystals have regular repeating units of atoms. When an X-ray beam passes through the crystal, the ray will interact with the atoms and scatter, leaving a pattern of dark marks in the film behind. The patterned marks constitute the photograph, and their shape and density provide clues to the arrangement of the atoms in the crystal – in other words, clues to the crystal's structure.

X-ray crystallography was used to decipher the structure of DNA because DNA can also form crystals. Figure 2.4 was named Photo 51 because it was the 51st photograph that Franklin and Gosling took, and the 51st time was the charm. The photograph was extraordinarily clear, the clearest picture of life's building block that we had at the time. When fellow scientists James Watson (1928–) and Francis Crick (1916–2004) saw the photograph, they immediately recognized its value and used it, together with other evidence, to derive that DNA has a double-helix structure – and received the Nobel Prize for it.

But, and here is the big but, does the photo tell *you* anything, even as you now know that DNA has a double-helix structure? I'll be the first to admit that it does absolutely nothing for me. I don't see a helix, let alone a double helix, and I have a hard time imagining how it *could* represent a double helix. All I see is an X pattern, with the center of the X being darker than its outer edges. I suspect that, to most readers, that's all they see too. It takes someone with significant and specialized training to deduce atomic structures based on X-ray crystallography. It also goes without saying that, merely by looking at Photo 51, readers won't be able to deduce or understand the scientific method used to obtain it.

Some may say that Photo 51 isn't a fair example because it isn't exactly a conventional photograph, one that was taken with an everyday camera. In that case, let's consider Figure 2.5, which *is* a conventional photograph.

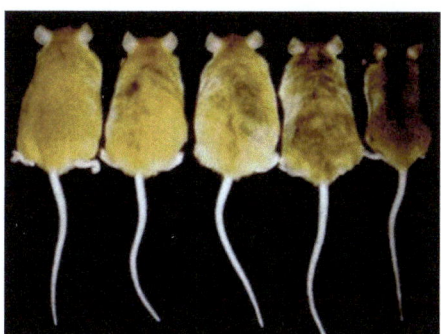

**Figure 2.5**   Different fur colors in mice as a result of their mothers' diets.

This photograph appeared in a *Science News* article by Brownlee on epigenetics, one of the newer research avenues in genetics. "Epi" means "on or above," so "epigenetics" literally means operating above the genetics. It studies how external factors such as the environment, diet, and behavior affect the way our DNA and genes (which are functional segments of DNA) work.

These external factors do not change the underlying DNA sequences. Instead, they modify the chemical groups that attach to DNA. Or they modify the proteins that DNA wraps itself around inside the cell nucleus. These modifications in turn determine whether a gene is turned "on" or "off." A common example of epigenetic influence is DNA methylation. DNA methylation means that a small chemical group called a methyl group (which consists of one carbon atom and three hydrogen atoms) is attached to DNA. When this happens to a gene, that gene is turned off and exerts no function. Studies of epigenetics can shed light on disease development and human health beyond the narrow focus on DNA sequences. For example, we know that genes that protect us from cancer may be silenced by epigenetic changes.

Figure 2.5 shows an instance of the epigenetic effect. All the mice in the photograph carry the same gene, called *agouti*, which creates brindle fur color that ranges from yellow to brown. But epigenetic factors – in this case the diet of the mice's pregnant mothers – can affect the eventual fur color. Mothers that were fed a soy-rich diet produced offspring that were more on the brown side of the color spectrum. The color change was caused by an ingredient in the soy diet – genistein – that added methyl groups to the *agouti* gene, thereby silencing its effect.

The photograph was originally published in an *Environmental Health Perspective* journal article and intended for an audience of trained scientists. In the journal article it was accompanied by 10 other visuals: graphs that show the statistical distribution of fur colors, graphs of methylation levels, diagrams, and DNA sequence letters showing methylation sites. When the study was reported to the general public in *Science News*, the photograph was the only visual reprinted.

At first glance, this would seem a good choice because, compared with the other, highly technical visuals in the original article, the photograph seems easy to understand. We can see five mice with different fur colors, corroborating the results of the study. In reality, however, the photograph alone tells us very little about *how* researchers got to these mice. The way diets affected the *agouti* gene, the inner working mechanism of methylation, the nature of the soy diet – these are not things that can be photographed.

Moreover, this photograph alone tells us very little about the significance of the original research. The research wasn't merely trying to show that diet affects mice's fur color. Rather, its most important finding is that a soy-rich diet, comparable to the human consumption level among some Asian populations, reduced the mouse offspring's weight. The brownest mouse in Figure 2.5 is noticeably smaller and thinner. A soy-rich diet, researchers therefore believe, could provide a favorable prenatal environment to protect humans from adult obesity and chronic diseases such as cancer. This significance is entirely lost in the photograph proper.

By this, I'm not suggesting that photographs like Figure 2.5 have zero value in popular science communication. They exhibit the visible outcome of research, and, paired with proper verbal explanation, they support the meaning of that research. What I am suggesting is that as life sciences change their focus, photographs, which only show external appearances, become less revealing. If people who are charged with popular science communication – be they scientists or science writers – do not recognize this and choose photographs as their default visual evidence, they risk overestimating the communicative power of their work.

Another caveat I should add is that the kind of record-keeping photographs commonly seen in early life science research did not become extinct. Photographs' cataloging value is still relevant. In ornithology, the branch of zoology devoted to studying birds, for example, we rely on photographs to identify the tens of thousands of bird species that inhabit the Earth. The same is true with ichthyology, the branch of zoology devoted to studying fish.

The National Oceanic and Atmospheric Administration's (NOAA) online gallery, for example, contains endless such examples that document diverse marine lives from fish to whales to birds. Figure 2.6 shows the appearance of a bigtooth cardinalfish (*Apogon affinis*). From its overall body shape and color to the detailed structures of its fins and eyes, Figure 2.6 demonstrates what a bigtooth cardinalfish looks like better than any other visuals or texts.

**Figure 2.6** Bigtooth cardinalfish (Apogon affinis).

## Photography as Human Intervention

The human intervention in photographic technology is a little difficult to conceptualize through modern cameras – those in our phones, for example. These cameras are completely sealed off and work their magic like blackboxes. All we see are lenses and screens. We press a button, and, voila, photographs appear, an automatic process.

The brief history of photography we saw at the start of this chapter, by contrast, should help to highlight just how much human intervention goes into photography. Early photographers were, literally, hands on. They messed with plates, experimented with chemicals, and traveled with portable darkrooms to process plates before they dried. Early photographers were also artists. They manipulated light, shadow, even retouched negatives to create aesthetic qualities in their work.

The development of color photography is especially telling. While the demand for color started almost immediately after the invention of photography, color photography, as mentioned above, didn't appear until much later. Eager to give the consumers what they wanted, photographers resorted to handing their black-and-white photos to artists for hand coloring.

Even with the invention of autochrome, which, as the name suggests, made color (*chrome*) automatically, the actual process was far from being mechanically objective. Notably, autochrome was a success because people found autochrome photographs beautiful. The photographs had a fuzzy effect reminiscent of impressionist paintings. This effect was created, unintentionally, by the dyed potato starch grains. In theory, the tiny, invisible grains would be randomly and evenly distributed to render color. But in reality, buildups were inevitable. These buildups were visible to the human eye, giving the photographs a misty, dreamy appearance. So, just as people applauded the ability of autochrome to record reality, they also admired its ability to transcend and beautify reality.

Taking apart early photography this way, hopefully, allows us to think of modern cameras in a similar light. In many ways, modern cameras afford us even more opportunities at intervention. Think about it – by changing the lens, exposure time, color scheme, photo processing software, etc., the same scene can look very different. Then there are higher-level human decisions such as which scenes to photograph, what position the photographer takes vis-à-vis the subject, and which photographs one eventually chooses to publish. Anyone who has taken photographs to share on social media will appreciate taking multiple pictures of the same subject and then choosing the best to share. You don't think that scientists do the same when photographing their research, or science writers do the same when picking images for publication?

Consider, for example, wildlife photography. Animals spend a lot of time sleeping, lying down, or doing things that we humans don't necessarily deem interesting or engaging. One has to wait a *long time* to get a shot of a whale jumping out of the water or an alligator eating prey. Inevitably, though, it is these interesting, engaging shots that are published and shared, creating a tampered impression of how animals act in the wild.

With this understanding of photographs as products of human intervention, we can now better appreciate – and question – the symbolic function of contemporary life science photographs.

## Contemporary Life Science Photographs as Symbolic Expression

What value photographs lost as direct scientific records and evidence in contemporary popular science communication, they gained in symbolic expression. That is, photographs are increasingly used as a way to evoke

**Figure 2.7** Picture of an elephant seal in NOAA's Monterey Bay National Marine Sanctuary.

public emotions and visceral reactions. They are vibrant in color, sharp in focus, and superior in composition. They showcase the precious and awe-inspiring life surrounding us.

Take Figure 2.7, a photograph of an elephant seal. This photograph, like that of the bigtooth cardinalfish shown in Figure 2.6, is published by NOAA. The elephant seal resides in NOAA's Monterey Bay National Marine Sanctuary. Despite this shared authorship, it is obvious that the two photographs are vastly different in function. Figure 2.6 provides an observational record and helps one to identify a bigtooth cardinalfish. Toward that goal, the photograph captures the entire body of the fish, head to tail, with no structural details obscured.

Figure 2.7, by contrast, seems not at all concerned about providing an observational record or helping readers to identify an elephant seal. Rather than photographing the entire body of the animal and its landmark features, the photograph zeros in on the animal's face. What this choice readily accomplishes is anthropomorphizing – in other words, humanizing – the elephant seal.

A close-up shot of the face is about the best way to humanize a nonhuman animal. In these shots, animals' facial structures and expressions are not all that different from our own. In fact, sometimes, with their round heads, big eyes, and an innocent stare, these animals easily remind us of human babies. No wonder that these photographs fill us with a feeling of connection and empathy. And that is precisely what Figure 2.7 intends to do – so that the public can appreciate marine lives as cute, precious, and worthy of our protection. The original caption of the photograph dovetails with this message: "Hello, I am here to interview for the role of 'cutest elephant seal' in NOAA's Monterey Bay National Marine Sanctuary! Will there be squid provided?"

The angles of the two photographs are no accident either. In Figure 2.6, the bigtooth cardinalfish is pictured in profile. The animal looks away from the viewer, forming what Kress and van Leeuwen call an "offer" photograph. That is, the photographed subject is offered to the viewer "as items of information, objects of contemplation, impersonally, as though they were specimens in a display case" (p. 119). In this kind of photograph, the viewer feels no direct contact with the subject that is photographed and feels no involvement with what the subject is doing.

By contrast, Figure 2.7 takes the frontal angle so that the elephant seal looks straight at the viewer. This is what Kress and van Leeuwen call a "demand" photograph. The photographed subject, through its gaze, "demands something from the viewer, demands that the viewer enter into some kind of imaginary relation with him or her" (p. 118). What kind of relation is demanded depends on the photograph, especially the facial expression of the photographed subject. A smile, for example, invites the viewer to enter into a friendly relation; a stern look invites the opposite. In Figure 2.7, the elephant seal's child-like cuteness and chubbiness, with a touch of helplessness, invites the maternal and paternal instincts in us to protect them.

Aside from anthropomorphism, another photographical trope favored by contemporary popular science is, for the lack of a better term, visual impact, achieved through fluorescent colors, impeccable lighting, striking compositions, and unusual subjects. These photographs showcase the mysterious, the exotic, and the fantastic life on Earth.

Take, for example, the Vizzies Visualization Challenges, an annual competition co-sponsored by the National Science Foundation and *Popular Science*.

Vizzies has three general judging criteria. "Visual impact" is the first, followed by "effective communication" and "freshness and originality." Wellcome Trust's Wellcome Image Awards (now rebranded Wellcome Photography Prize) is another prestigious competition. It too has three judging criteria. The first is "creativity," which is defined along the line of visual impact, followed by "storytelling" and "technical."

Previous winners of these competitions include a photograph of a chameleon that had been dipped in chemicals to make its skin and muscle transparent, processed with enzymes to digest its flesh, and colored with dyes to highlight its bones and joints. The resultant photograph is a stunning transparent chameleon with purple bones and blue joints – it's a wow.

Figure 2.8 is also a previous winner. It shows a preserved horse uterus, about five months into pregnancy, with the fetus removed but still attached. The specimen had been preserved in formalin for 40 years and is stored in the

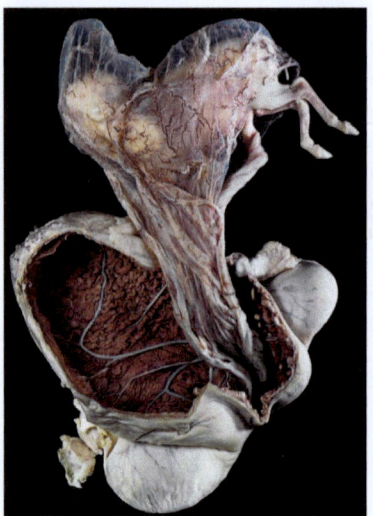

**Figure 2.8** Award-winning photograph of a horse uterus with the fetus removed but still attached.

Royal Veterinary College's Lanyon Anatomy Museum. With its striking composition, unusual colors, and gritty details, the photograph is, once again, visually stunning (if also a little graphic).

What functions do these visually striking photographs perform? What effects do they have on public viewers? In many ways, these photographs *are* about science, but at the same time they are not concerned about scientific research *per se*. I like to think of them as symbolic expressions of science. They elicit awe and fascination from viewers about the photographs, the photographed, and, by extension, about science. This reaction can inspire public interest in and engagement with science, encouraging viewers to visit science museums, read science books, even pursue scientific studies – all of which are goals of popular science communication.

At the same time, awe and fascination will do good for the enterprise of science. An awestruck and fascinated public is more likely to support scientific efforts and government spending on science, and to trust science and scientists, which will in turn lead to more scientific efforts and spending. I point this out not as a criticism, only as a reminder that science is not a disinterested pursuit that happens in a vacuum. It is a large social enterprise that employs millions of people and runs on billions of dollars. If nothing else, this alone should guarantee that no science photographs will be 100% mechanically objective.

## Contemporary Life Science Photographs as Symbolic Excess

If science photographs can be used as symbolic expressions to elicit emotions and reactions, is there a risk of going too far? Well, you tell me.

In 2004, New Zealand activist group Mothers Against Genetic Engineering in Food and the Environment (MAdGE) put up a billboard in Auckland and Wellington. Pictured on the billboard, as Bloomfield and Doolin described, is "a photographic image of a naked, four-breasted young woman, kneeling on all fours in side profile, with her breasts hooked up to a dairy milking apparatus and a red 'GE' brand on her buttock. The accompanying press release was titled 'Why Not Just Genetically Engineer Women for Milk?'" (p. 515).

Clearly, the photograph is not of a real woman. It is a result of photographic editing, using software such as Photoshop. Still, the effect was very realistic. The photograph, obviously, is not meant to function as a piece of scientific evidence, but to provoke public attention, debate, and outrage. According to MAdGE, the billboard is designed to protest against genetically modified organisms (GMOs), in particular against inserting human genes into cows to create designer milk. Because the photograph's emotional content outpaces its material reality, I call it, after Bloomfield and Doolin, a symbolic excess.

What is one to think and feel about the photograph? At first glance, it is shocking, offensive even. The photographed young woman is completely naked, creating an erotic, voyeuristic display where the female body is objectified for inspection. At the same time, that very shock factor may be what is needed to shake things up, to shock people out of their complacency about GMOs and to ask hard questions about GMOs' ethical and social impacts.

Then again, is a shocking photograph like this going to help promote useful, fine-grained discussions about GMOs? In purposefully building strong visceral reactions to offend, the photograph can easily create division. People will either "get" the message and align themselves with MAdGE – a mother or mother-like figure who, presumably by nature of their nurturing experience and instinct, deeply understands the dire consequences of GMOs – or people will reject the message as doctored, sensationalized, and offensive, and identify themselves on the opposite side of MAdGE as a rational, informed citizen who (perhaps) supports genetic research.

Meanwhile, finer details surrounding GMOs – in this case, genetically modified dairy milk – fade into the background. For example, dairy milk contains the $\beta$-lactoglobulin protein, which causes allergic reactions in 2–3% of human infants. Genetically modified dairy absent of $\beta$-lactoglobulin is now a reality, as reported by Jabed et al. in *Proceedings of the National Academy of Sciences*. For human mothers who cannot breastfeed, this may come as a relief. On the other hand, the genetically engineered cow was born without a tail. Jabed et al. acknowledged that this was a rare congenital abnormality and speculated that it was not caused by the suppression of the $\beta$-lactoglobulin protein. The actual cause for the deformity and its potential impact must nevertheless be clarified.

**Figure 2.9**    A photograph portraying the genetic modification of a tomato.

Granted, a naked woman with four breasts is an extreme example. We are not likely to see photographs like that in mainstream popular science publications or the mass media. Yet, photographs that are similar in function do exist. For example, an ear of corn might have its husks peeled back to reveal not yellow kernels but colorful pills, implying the genetic engineering of corn for medicinal value. Or succulent vegetables and fruits might take a shot of suspicious fluid from a syringe to imitate the process of genetic engineering, such as shown in Figure 2.9.

These photographs are much less shocking than those of naked, four-breasted women, but their symbolic functions are similarly complex and excessive.

On one hand, it is possible to see them as a clever figure of speech to illustrate unfamiliar concepts. The molecular-level differences between GMOs and non-GMOs are not visible to the naked eye and can't be revealed through photographs. The process of creating GMOs is complicated and not conducible to a simple photograph either. Therefore, the creators of these photographs took a metaphorical approach. Color pills took the place of altered genes, and physical injection of fluids replaced genetic engineering.

On the other hand, it is possible to see these photographs as a form of misuse, a "bad" metaphor, if you will, because what is photographed is more than what the material reality warrants. A common approach to creating GMOs starts with transferring a target gene into bacteria, which in turn slip the gene into a plant cell. After taking up the gene, the cell divides, grows into a seedling, becomes a vegetable or bears fruit. This and other genetic engineering processes bear no resemblance to injecting fluids into harvested food.

The potential effect of these photographs on public readers is similarly complex. In creating jarring scenes, these photographs have the ability to catch people's attention and encourage them to question science and its media allies' narrative about genetic engineering. As figures of speech, they can function as entry points and create moments of dialog and engagement. For example, some people may wonder, "Is this *really* how we produce a genetically modified tomato?"

The risk, similar to photographing naked, four-breasted woman, is that the emotions elicited by the photographs will override abstract thinking and meaningful conversation. The photographs can become short-hand communication – Frankenstein food! – that quickly rallies likeminded people on a moral ground and turns off people who don't share that sentiment. When the conversation is at the level of Frankensteining, we lose opportunities to discuss GMOs' potential to relieve world hunger or their risk to the ecosystem.

## The Line Is Blurred

Although I presented early life science photographs as objective evidence and contemporary ones as symbolic artifacts, it is important to note that the difference between the two isn't clear-cut. It is possible to say that a photograph is primarily created to function as one or the other, but frequently traces of both can be found in the same photograph.

Anna Atkins, for example, was praised for her artistic sensitivity and for arranging algae in imaginative and elegant ways while photographing them. Duchenne de Boulogne intended his catalog of human facial expressions to guide sculptors and actors in their artistic portrayals of the human soul. The fractured skulls in *Photographs of Surgical Cases and Specimens* are reminders of the human cost of war. And if the skull, clinically pristine and devoid of

**Figure 2.10** Double amputation of the forearms for injury caused by the premature explosion of a gun.

human context, doesn't seem symbolic enough, then we have plenty of human portraits in the same book.

Photographed in Figure 2.10 is Private Samuel H. Decker, who was wounded during the 1862 battle of Perryville, Kentucky. According to the case narrative, while ramming his gun, Decker had "half of his right forearm, and somewhat less of the left, blown off by the premature explosion of the gun" (volume 5, photograph no. 205). During the Civil War, arm and leg wounds were often treated with amputation because the wounds were contaminated by splintered bone, dirt, and torn clothes that were time-consuming and difficult to clean. Risk of deadly infection left surgeons with no choice but to amputate.

In Decker's case, both forearms were amputated 5 hours after the accident and, about three months later, the wounds had completely healed. Admirably, after his recovery, "Mr. Decker began to make experiments for providing himself with artificial limbs. He produced, in March, 1865, an apparatus hitherto unrivaled for its ingenuity and utility" (volume 5, photograph no. 205). The artificial limbs pictured in Figure 2.10 were, presumably, Decker's own inventions.

Decker "receives a pension of $300.00 per year, and is a doorkeeper at the House of Representatives ... With the aid of his ingenious apparatus he is enabled to write legibly, to pick up any small objects, a pin for example, to carry packages of ordinary weight, to feed and clothe himself, and in one or two instances of disorder in the Congressional gallery has proved himself a formidable police officer" (volume 5, photograph no. 205).

Aside from providing detailed records of injuries, surgeries, and surgical outcomes, photographs like Figure 2.10 are records of the tragic human cost of war. Or, perhaps, in some viewers, they also elicit an admiration of heroism, even a romantic notion of war. Noticeably, Figure 2.10, as other photographs of wounded and surviving soldiers in *Photographs of Surgical Cases and Specimens*, assumes the frontal angle. As mentioned above, this angle creates a demand photograph that invites viewer reaction and involvement.

Indeed, whenever human bodies are photographed in a scientific context, they are bound to stir up visceral reactions. This is why gross anatomy photographs are used in scientific journals and dissection classrooms but rarely in popular science communication. Although these photographs are undeniable evidence of surgical techniques and human anatomies, they are considered too graphic for public consumption.

When they were dangled in front of public eyes, especially in a purposefully sensational format, the line between respectable science and vulgar spectacle became hard to discern. Figure 2.11, dated 1911, was created by the controversial French surgeon Eugène-Louis Doyen (1859–1916) using the so-called topographical technique. The technique mummifies human bodies and then slices them up layer by layer to show the changing, underlying anatomies. Notwithstanding the scientific

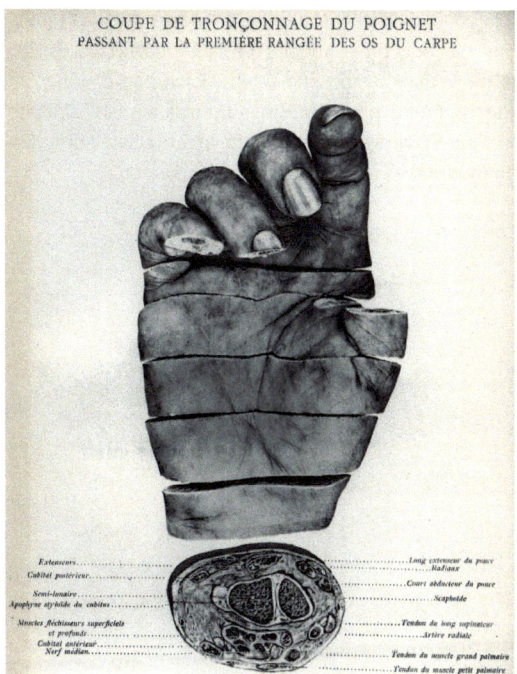

COUPE DE TRONÇONNAGE DU POIGNET
PASSANT PAR LA PREMIÈRE RANGÉE DES OS DU CARPE

**Figure 2.11** Photograph of the right hand, by *Eugène-Louis Doyen*.

annotations, the photograph looks, to me at least, more horror than science. What does it look like to you?

## Conclusion

Photographs play multiple roles in the communication of life sciences. They can be used as scientific evidence to convey methods and findings; they can be used as symbolic artifacts to incite sympathy and outrage. They can become a tool to engage and initiate conversations, but also a tool to divide and shut conversations down. Sometimes, what you see in them is what you get, but often, especially in contemporary life science communication, that is not the case.

These complexities, I hope, demonstrate that photographs are not at all "easy." To a public audience they may not convey the proverbial thousand words. Or, indeed, they belie a lot more than a thousand words. Everyone involved in creating and consuming scientific photographs – from scientists, science writers, photographers, to public viewers – needs to be aware of the complexity of this seemingly commonplace visual and its potential effect.

# 3 Micrographs

Micrographs, like the little (pun intended) cousin of photographs, are considered by some as an objective portrayal of nature. Why, they are photographs of the microscopic world invisible to the naked human eye. As such, what you see is what you get, and what you get is nature unveiled.

Particularly because the microscopic world is invisible to us in everyday life, we find it even more urgent to behold that world. We assume that if and when we see, we will automatically understand. If and when we observe microorganisms in their smallest components, we will be able to "get" them and conquer them.

This chapter argues that these are misconceived or oversimplified perceptions of micrographs.

## A Brief History of Micrography

Micrographs are also known as photomicrographs. Simply put, they are photographs taken through a microscope to show an enlarged view of a small object. To attempt to describe the history of micrographs, then, we must start with the history of microscopes.

Embarrassingly enough, that description starts with the admission that we don't know who, exactly, invented the first microscope. Magnifying lenses have existed for centuries and were used in artisans' workshops. As far as concrete record goes, many credit Dutch spectacle-maker Zacharias Janssen (1580–1638), with the help of his father, for having made the earliest microscope in the 1590s. Janssen's microscope used multiple lenses to bend light and enlarge samples to 3–9 times their original sizes.

Three to nine times magnification isn't going to cut it if one intends to perform serious scientific observations. Plus, early microscope designs suffered from unfocused lighting and poor image quality. It took another Dutchman, Antonie van Leeuwenhoek (1632–1723), to make the scientific application of microscopes a real possibility. In the 1660s, Leeuwenhoek invented single-lens microscopes, which could better focus light and achieve magnifying power of up to 270 times. Using his superior devices, Leeuwenhoek studied the microscopic structures of seeds, fish scales, nerves, muscle fibers, and more. In fact, he is credited with the discovery of bacteria, among other miniature things.

These early microscopes are the origin of today's light microscopes (also known as optical microscopes), so called because they rely on lenses and light to magnify objects. Early light microscopes use external light sources such as oil lamps and candles, which do not provide reliable illumination. By contrast, modern light microscopes have built-in light sources, often a bulb or laser, that can be easily controlled.

Early devices were not equipped with cameras to take a picture of the observation. As mentioned in Chapter 2, the earliest photograph didn't come into existence until the nineteenth century. To record what they saw under the microscope, the early microscopists had to manually draw the sight. The most famous such drawings are, without doubt, those in Robert Hooke's 1665 *Micrographia*. We met this masterpiece in Chapter 1, via the giant flea, and here it is again, a giant louse clutching a human hair (Figure 3.1).

Almost two centuries would pass before we see actual micrographs. English inventor Thomas Wedgwood (1771–1805), who was, by the way, Charles Darwin's uncle, proposed the idea of micrographs in an 1802 paper published in the *Journals of the Royal Institution*. In it, he stated that "I have found that the image of small objects, produced by means of the solar microscope, may be copied without difficulty on prepared paper ... it is necessary that the paper be placed at but a small distance from the lens."

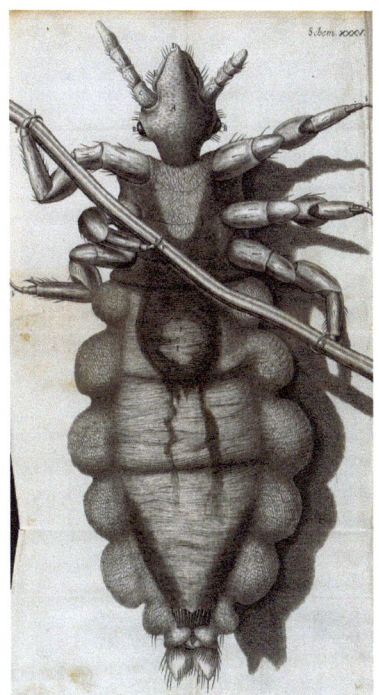

**Figure 3.1**    Robert Hooke's depiction of a louse.

A solar microscope, as the name suggests, uses the sun as the light source. The device is fitted into the window shutter of a darkened room. One end of the device, which has a mirror, is placed outside the room to collect sunlight. The collected sunlight is condensed by a lens to illuminate a sample and then projects an enlarged image of the sample onto the opposite wall. This image can be viewed by a large audience at once, making for a great "science party" for the genteel society of the nineteenth century. When the projected image is photographed, there you have it: a micrograph in its earliest and simplest form.

There is, however, no evidence that Wedgwood actually turned his theories into practice, as no micrograph from him survived. Instead, English scientist

**Figure 3.2**   William Henry Fox Talbot's micrograph showing the cross section of a stem.

and photography pioneer William Henry Fox Talbot (1800–1877) left behind one of the earliest, if not the earliest, solar micrographs. Dated 1839, it shows the cross section of a stem (Figure 3.2).

Today, solar microscopes have retired into museums. Light microscopes, though, remain widely used in life science research. In addition to internal light sources, various other advances have happened along the way, from the phase contrast technique that enhances contrast in transparent samples, to fluorescent dyes that create precise color coding of different structures in one sample. Modern light microscopes also come equipped with digital cameras that can easily capture high-resolution color photographs, even projecting them to a screen for easy viewing.

Aside from light microscopes, we now also have electron microscopes. These devices use beams of accelerated electrons as sources of illumination. Electron microscopes have higher resolutions than light microscopes, because the wavelength of electrons is much shorter than that of photons.

Different types of electron microscopes exist. Two common ones are transmission electron microscopes and scanning electron microscopes. A transmission electron microscope works by passing (aka transmitting) electrons through an ultrathin sample and using the electrons that pass through to create an image. Because the electrons pass through the sample, we get to observe the internal structure of the sample. These microscopes are capable of magnifying samples by more than 50 million times.

A scanning electron microscope, by contrast, scans the surface of a sample with an electron beam and uses the electrons that are bounced back to generate an image. In other words, it helps us to see the sample's external structure. Scanning electron microscopes have smaller magnification power, capable of magnifying samples by only 1–2 million times. They can, however, scan a larger area and provide better depth perception, so the resultant micrographs take on a three-dimensional appearance, as we will see later.

## Seventeenth-Century Microscopic Drawings: Observed Imagination

Early microscopes had limited magnification power and image resolution. They suffered from aberration (the deviation of light rays) and insufficient lighting, which led to blurred and dark images. Robert Hooke, the creator of the giant flea and louse images, had to put up with these limitations.

Yet, you would never know this by looking at his microscopic drawings. Whether it was plant cells or giant insects, the drawings were extremely detailed, meticulous, and realistic. One can't help but wonder how much of that effect was due to Hooke's observation and how much was his imagination.

Short of traveling back in time and looking over Hooke's shoulder, we can't have definitive answers. What we can do is look at modern-day micrographs of the same critters, the head louse for example (Figure 3.3), to gain some perspectives. Figure 3.3(a) is a view of the female head louse lying belly-up. Figure 3.3(b) and (c) respectively, are views of the female and male head louse lying belly-down.

If we compare these micrographs with Hooke's drawing, it is clear that Hooke drew the view shown in Figure 3.3(a). At a glance, one immediately marvels at the amount of similarities between the two: from the beady eyes to the hooked feet, the drawing captures all the structural details of the critter. In fact, the drawing seems "more real" than the micrograph. It gives the louse more dimension by using shadows. It provides more definitive details: a shiny, armor-like texture, especially in the legs; an almond-shaped structure with

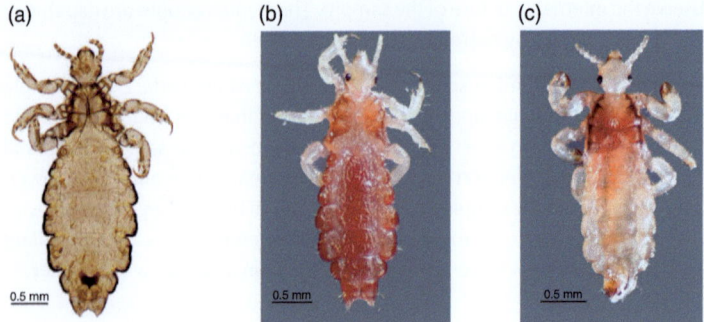

(a)    (b)    (c)

**Figure 3.3**    Contemporary micrographs of the louse.

trailing "ribbons" in the stomach. The drawing also gives the louse more character, more presence, so to speak: its body is shaded with intricate patterns and a sense of perfection not evident in the modern micrograph; even the placement of bristles seems well controlled and delicately balanced. The louse, despite being a louse, looks positively regal in Hooke's drawing.

Given the microscope available to Hooke, which would be far inferior to the one used to create Figure 3.3, it is doubtful that Hooke physically observed *all* the details he drew. Yes, he undoubtedly made detailed observations, but he also used his creative license to fill in the gaps, if only to make the animal more spirited, more perfect.

This is especially plausible, and celebrated, when we consider that Hooke lived in an era that adored visuals that transcend nature to unveil God's intention (see Chapter 1). To reveal the true nature, one needs genius imagination.

## Nineteenth-Century Microscopic Drawings: Tension between Science and Art

Micrographs became a technical possibility in the early nineteenth century. Right around the same time arose the obsession with mechanical objectivity in scientific visuals (see Chapter 1). As German bacteriologists Carl Fraenkel (1861–1915) and Richard Pfeiffer (1858–1945) put it (quoted by Daston and

Galison), "A drawing can only be the expression of a subjective perception and therefore must, from the beginning, renounce the possibility of an objection-free reliability . . . The photographic plate, by contrast, reflects things with an inflexible objectivity as they really are, and what appears on the plate can be looked upon as the surest document of the actual conditions" (p. 177).

According to Fraenkel and Pfeiffer, we see not only with our eyes but also with our minds. In other words, we see what we *want* to see and filter out things we don't want to see as "unimportant." A drawing, then, by definition can't be objective. Micrographs, by contrast, see everything, desirable and undesirable. They alone stand a chance to objectively capture the microscopic world.

This belief, of course, is oversimplified, as we will see later. And regardless, microscopic drawings didn't simply disappear. Part of this was because of the lingering truth-to-nature aesthetics. Part of it was because of the technical limitation of nineteenth-century micrographs. Yes, micrographs could capture many optical details, but they would also, as Daston and Galison wrote, capture optical interferences such as the impurities of a sample, defects of a plate, or dust particles. Sometimes, details would vanish when there was too much light or too little light. In addition, micrographs could capture only a small part of the sample and had a limited focal point, so at the edge the visuals became blurred.

By contrast, when drawing, an observer could filter out optical interference, could move the sample from side to side to inspect every corner, and could provide a better depiction of depth. So it is that in the nineteenth century we still see nature as perfection in micrographic drawings. The most celebrated (and also controversial) are those created by German biologist Ernst Haeckel (1834–1919).

Pictured in Figure 3.4 is a plate from Haeckel's 1862 masterpiece *Die radiolarien: Eine monographie* (*The Radiolarians: A Monograph*). The book, as its title suggests, is dedicated to radiolarians, which are single-celled microorganisms that live in the ocean, from the water surface to the deep sea. Included in Figure 3.4 are a few among the thousands of radiolarian species described in the book and elsewhere by Haeckel. With their perfect circles, intricate patterns, and delicate symmetry, these radiolarians are nature enhanced and idealized.

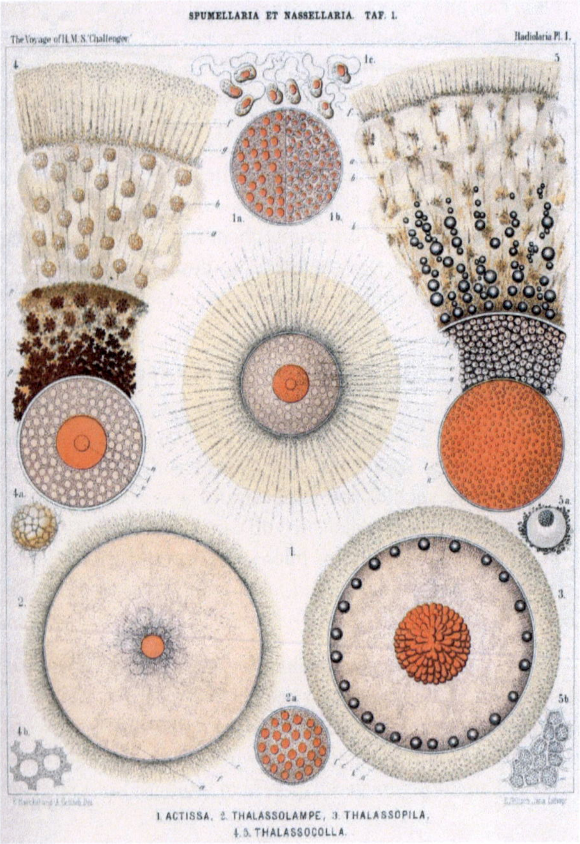

**Figure 3.4**    Ernst Haeckel's drawing of radiolarians.

Today's viewers have no problem calling these microscopic drawings art. Indeed, Haeckel's radiolarians are sold as art and decors, printed on canvases, greeting cards, and T-shirts. Had these drawings happened a few centuries back, Robert Hooke and his contemporaries would have no problem seeing them as the magnificent truth of nature.

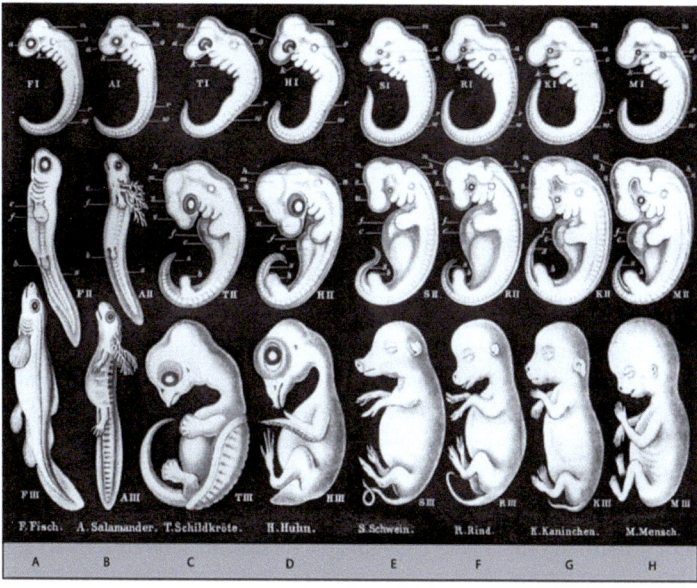

**Figure 3.5** Ernst Haeckel's drawing of embryos.

But, during Haeckel's time, with the pursuit of mechanical objectivity on the rise, his stylized drawings made him seem peculiar and eccentric. Particularly controversial was Haeckel's drawing of animal embryos. Embryos at the earliest stages of development are tiny and require a microscope for detailed observation. In the 1874 edition of his book *Anthropogenie: oder, Entwickelungsgeschichte des Menschen* (*Anthropogenie: Or the Developmental History of Human*), Haeckel drew different animals, from fish to tortoises to pigs to humans, going through almost identical stages of early embryonic development (Figure 3.5). Haeckel used these visual similarities to support his belief that a species' embryonic stages represent the adult forms of its ancestors. In other words, for "higher" animals such as humans, their embryos recap the adult forms of "lower" animals such as fish. The "higher" animals would go past those early forms and keep evolving into what they are.

We now know that Haeckel's theory of evolution is incorrect. "Higher" animal embryos don't pass through the adult forms of "lower" animals. Indeed, evolution isn't progressive so there are no higher or lower animals. That said, embryos of related species do display similarities during development, which supports the Darwinian theory of evolution and common descent.

Without committing his belief to drawings, Haeckel would have simply been mistaken in his interpretation of data. But, *because of* his drawings, he was harshly criticized. His mistake, it was thought, wasn't an honest one.

Haeckel's contemporaries, such as German embryologist Wilhelm His (1831–1904), accused Haeckel's embryo drawings of being willful exaggerations, forgeries even. American geneticist Richard Benedict Goldschmidt (1878–1958) lamented that "Haeckel's easy hand at drawing made him improve on nature and put more into the illustration than he saw ... one had the impression that he first made a sketch from nature and then drew an ideal picture as he saw it in his mind" (quoted by Richardson et al. p. 104).

Most recently, in 1997, British embryologist Michael Richardson and his team studied some 40 different species' embryos. Based on their observations, Richardson argued that Haeckel's drawings both added and omitted features to exaggerate the similarities between different animal embryos. For example, the drawing of the early-stage human embryo lacks the "limb buds" (which will eventually develop into limbs). Or, the eye of the early-stage chick embryo is blackened to resemble those of a mammal. Richardson concluded that Haeckel's drawing was "one of the most famous fakes in biology."

Without doubt, there *are* deviations between Haeckel's drawings and microscopic observations. Part of this was due to the inferior microscopic instrument Haeckel had at his disposal and the rush to get the first edition of *Anthropogenie* to the printing press. As science historian Robert Richards pointed out, in the later editions of his book, Haeckel's embryo drawings contained more differentiating details between the animals.

But that's not all. By Haeckel's own admission (quoted by Daston and Galison), the drawings were not meant to be "exact and completely faithful ... but rather ... illustrations that show only the essentials of an object, leaving out inessentials" (p. 191). Moreover, as quoted in Hopwood, Haeckel showed the essentials not really as he *saw* it, but as he *thought* it (p. 161).

Does this constitute lies and frauds? To nineteenth-century – and today's – scientists who lean toward mechanical objectivity, probably yes. Microscopists must copy what they see, not elaborate on what they see. They must be a mere conduit, not a creator, much less an artist. To Haeckel, who believed in the importance of art, the importance of truth to nature, no. To mindlessly copy nature is to drive all ideas out of science, as he put it.

Some contemporary scholars, such as Watts et al., have come to Haeckel's defense, arguing that his drawings were not meant to be precise records or technical drawings, but simplified teaching tools for a general audience who needed help seeing the essentials in microscopic observations. Such drawings, they argue, are more instructive than actual micrographs.

It is not my place to speculate about Haeckel's intentions and actions; extensive work has been done on the subject without being able to say one way or the other. But that much is clear: microscopic drawings *are* more instructive than micrographs, as we will now see.

## Contemporary Microscopic Drawings: The Triumph of Interpretation

Figure 3.6 is a typical depiction of mitosis, aka nuclear or more broadly cell division, where one cell divides into two identical cells. The drawings are color-coded, a contemporary norm, to differentiate important components such as the chromosomes and the spindle fibers. Simple lines and shapes are used to denote minute details such as the thickening of the chromosomes or the elongation of the cells prior to division. Together, these techniques help to depict the multiple stages of mitosis and the corresponding cellular changes.

Even with the technical advances we have made in high-power microscopes and high-resolution micrographs, no micrograph will be able to rival such simplified, stylized drawings. The drawings will be sharper, clearer, and more defined, because the salient features have been interpreted for a viewer. This effect can be seen by comparing Figures 3.6 and 3.7, the latter being the corresponding micrographs of mitosis.

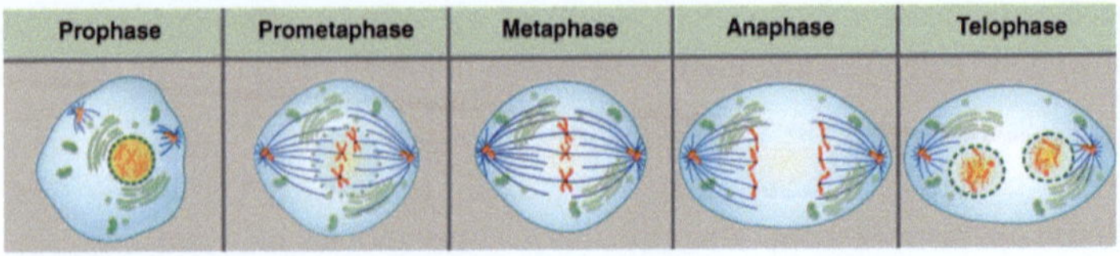

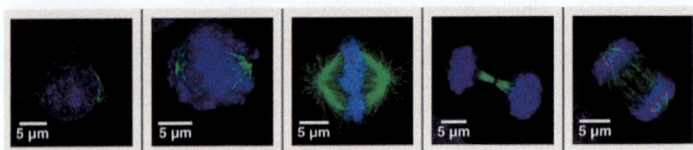

| Prophase | Prometaphase | Metaphase | Anaphase | Telophase |
|---|---|---|---|---|

**Figure 3.6**   Contemporary microscopic drawing of mitosis.

**Figure 3.7**   Contemporary micrographs of mitosis.

Contemporary readers who are used to seeing microscopic drawings such as Figure 3.6 will be surprised by how these drawings were self-consciously defended in the early twentieth century when interpreted visuals were just starting out. Figure 3.8 is a 1900 microscopic drawing of the tip of a growing onion root. The drawing shows much the same information as Figure 3.6, just not sequentially arranged.

The drawing was republished in a 1908 *Scientific American* magazine article. In the article, the author, M. A. Lane, repeatedly assured readers that despite the visual being a drawing, it was what the readers would have seen with their own eyes had they looked into the microscope: It is "an excellent picture of *the sight one sees* when one examines an extremely thin section cut from the tip of a growing onion root," the author wrote (p. 45, emphasis mine). "Here we *see* the cells bound together … Here and there in the drawing is *seen* a cell in which the nucleus is replaced" (p. 45, emphasis mine). One gets the impression that the author would very much like to use a micrograph but simply couldn't, likely because high-quality, high-resolution micrographs weren't available.

Such self-conscious reassurance has all but disappeared in today's use of microscopic drawings. The most we get is the combined use of drawings and

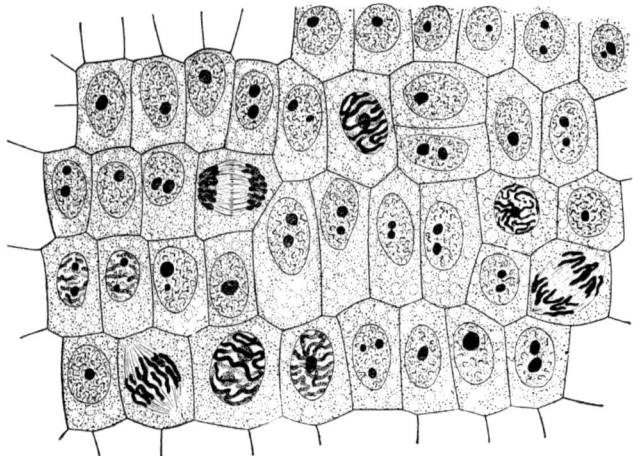

**Figure 3.8**   Microscopic drawing of the tip of a growing onion root.

micrographs (Figure 3.9). This combined view arguably provides the best of both worlds: a simplified interpretation as well as an opportunity to situate that interpretation in the microscopic vision. Readers who only want the quick lesson can get what they want; those who are curious to see the "authentic" view can get what they want, as authentic as a micrograph can get.

## Looking At vs. Looking For

In early 2020, at the start of the coronavirus outbreak, scientific journal *Nature* published a study by Zhou et al. on the identification and characterization of the novel virus SARS-CoV-2. According to the study, the virus uses the same way to enter and infect cells as the SARS-CoV-1 virus that caused the 2002–2004 SARS outbreak.

*Science News* picked up the story and reported it to the general public. Among the many visuals used in the original *Nature* article, *Science News* chose to republish one, a micrograph (Figure 3.10).

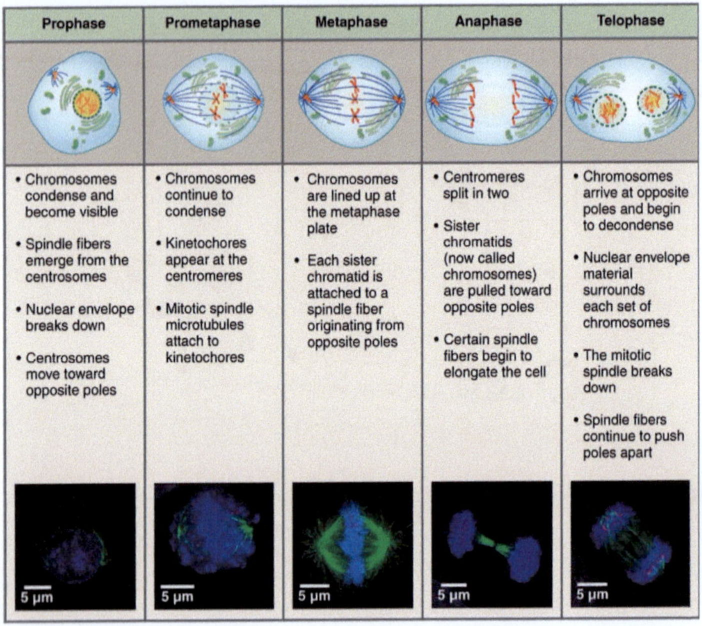

| Prophase | Prometaphase | Metaphase | Anaphase | Telophase |
|---|---|---|---|---|
| • Chromosomes condense and become visible<br><br>• Spindle fibers emerge from the centrosomes<br><br>• Nuclear envelope breaks down<br><br>• Centrosomes move toward opposite poles | • Chromosomes continue to condense<br><br>• Kinetochores appear at the centromeres<br><br>• Mitotic spindle microtubules attach to kinetochores | • Chromosomes are lined up at the metaphase plate<br><br>• Each sister chromatid is attached to a spindle fiber originating from opposite poles | • Centromeres split in two<br><br>• Sister chromatids (now called chromosomes) are pulled toward opposite poles<br><br>• Certain spindle fibers begin to elongate the cell | • Chromosomes arrive at opposite poles and begin to decondense<br><br>• Nuclear envelope material surrounds each set of chromosomes<br><br>• The mitotic spindle breaks down<br><br>• Spindle fibers continue to push poles apart |
| 5 µm | 5 µm | 5 µm | 5 µm | 5 µm |

**Figure 3.9**   Combined use of microscopic drawings and micrographs.

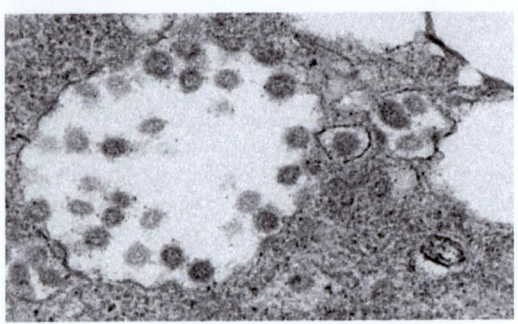

**Figure 3.10**   Electron micrograph showing the infection of cells by SARS-CoV-2.

As the *Science News* article writer Erin Garcia de Jesús explains, SARS-CoV-2 attaches itself to a protein (known as ACE2) that resides on the surface of a normal cell. Once attached, the virus can then poke its way into the cell, "essentially picking the cellular lock with a spiky protein on the virus's [own] surface."

This is an excellent metaphor, but the micrograph, prominently displayed with the article, fails to illustrate the idea. The caption to the micrograph, as used in the *Science News* article, reads "The new coronavirus (small circles shown in this electron micrograph) spreading in China and several other countries uses the same cellular protein as SARS to gain access to cells."

At the most basic level, one needs to identify the different components in the micrograph. There is a large circular area with a light background. Is it the cell? Or is the whole grayish area surrounding the circle the cell? Within the light circle are dozens of small, roundish dark spots. Are these dark spots the "small circles" and hence the viruses? Toward the middle right of the micrograph, one can make out a few more round, dark spots with black outlines. Are these the viruses instead, or are they viruses as well? And what is the outline surrounding them? Is it the cellular protein ACE2? There is an odd-shaped protrusion toward the top right of the micrograph. Is that ACE2 instead? In the bottom right corner, one sees another roundish, dark object. What is that? Everything is roundish and darkish, and a "look at X" micrograph can easily become a "look for X" exercise.

If the micrograph is colored, that would certainly help. Minimally, the cell and the virus can be in different colors. And indeed, contemporary micrographs are often presented in color. Either the samples are stained during observation, or the micrographs are digitally colored in post-processing (e.g., using Photoshop).

But color alone doesn't solve the problem. Microscopic objects have no counterpart in the visible, physical world, so there is little visual reference that nonspecialist viewers can use to try to comprehend a micrograph. Microscopic instruments and related conventions are not common knowledge, which makes a micrograph less intuitive. Bottom line: If one doesn't know what one is looking at, no amount of coloring is really going to help.

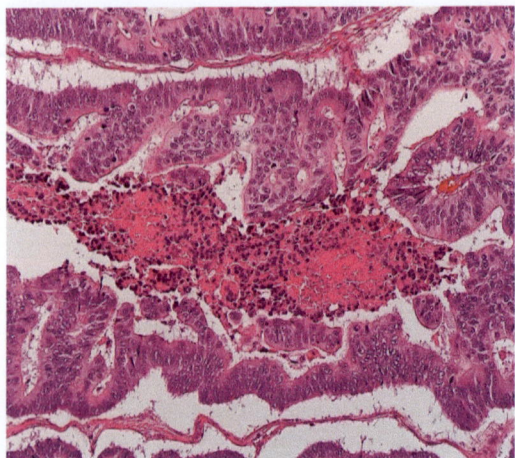

**Figure 3.11**   Micrograph of colorectal cancer.

Take, for example, Figure 3.11, which shows colorectal cancer. The micrograph is supposed to demonstrate moderately differentiated colorectal cancer with "dirty necrosis," which are malignant glands, cell debris, and inflammatory cells. For readers who do not know what cancer cells, glands, or inflammatory cells look like, this lavender–magenta–purple micrograph may as well be in black and white.

## Snapshots vs. Stories

Even when readers can locate the structure of interest in a micrograph, that doesn't mean that the micrograph will now automatically make sense. In Figure 3.10, for example, the micrograph does very little to support the gist of the *Science News* article: that the virus enters the cell by way of ACE2. The original *Nature* article that inspired the *Science News* article contained more supporting visuals. Singled out and standing on its own, Figure 3.10 is a mere snapshot, contextless and marginally helpful.

Figure 3.10 is not an isolated example. Browsing through popular science publications, one sees many snapshot micrographs. As another example, in an

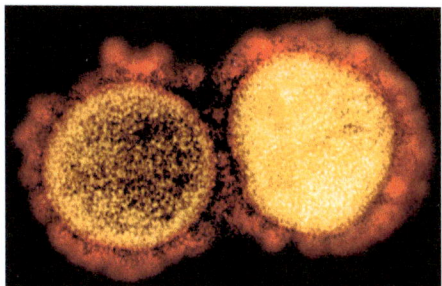

**Figure 3.12**    SARS-CoV-2 snapshot micrograph.

August 2021 NBC news article by Denise Chow, we are told that the delta variant of the novel coronavirus spreads more easily because it replicates faster and accumulates in higher amounts in infected patients. A single visual, a micrograph (Figure 3.12), accompanies the article. The caption to the micrograph reads "A transmission electron micrograph of SARS-CoV-2 virus particles."

The micrograph is not irrelevant, nor is it highly relevant. It is a snapshot that illustrates nothing of what the article is about. In fact, this micrograph was originally released by the National Institute of Allergy and Infectious Diseases back in March 2020, before the delta variant was even known.

If these snapshot micrographs aren't all that informative, why do we see so many of them? There are several possible reasons.

First, some of these micrographs are visually attractive, such as Figure 3.12 with its glowing red and yellow virus particles against a black background. These micrographs help to draw readers' attention (more about this function below).

Second, like with photographs, there is the misconception that micrographs are easier to understand than specialized life science visuals such as phylogenetic trees or sequence alignments. After all, the idea goes, micrographs capture what we see and so require no extra knowledge to understand.

Third, snapshot micrographs reflect our belief – obsession even – that if and when we can *see* (this invisible little thing), we will be able to understand. The more minutely and the more clearly we can see, the better. Obviously, this is true to some extent. But an obsession with seeing the invisible can become reductive when micrographs are taken out of context to zero in on nothing but *the thing*.

Last, these snapshot micrographs help assure readers that visual observations had been made: The novel coronavirus was *seen* to enter the cells a certain way; the delta variant was *seen* to accumulate rapidly in infected cells. In other words, these snapshot micrographs are doing the same rhetorical work that verbal reassurance did in the 1908 onion root article. In the absence of a high-quality micrograph, the writer of the onion root article had to verbally assure readers of visual observation. With high-quality micrographs available, contemporary writers moved swiftly on.

Yet, public distrust of science, which the COVID-19 outbreak abundantly highlighted, is not going to be curtailed by the mere presence of some snapshot micrographs. To help bring about trust, understanding, and engagement, we need micrographs that constitute accessible and meaningful evidence.

One way to do this is by telling stories through micrographs. The series of micrographs in Figures 3.7 and 3.9, for example, show the progressing changes that happen during cell division. The change constitutes the story, which constitutes the scientific evidence in question. Even if readers cannot pinpoint each individual structure in each micrograph, they can see the changes that happen between micrographs.

As another example, Zhou et al.'s *Nature* article, the one behind the *Science News* article, used a series of micrographs to show that cells with the ACE2 protein were infected by the SARS-CoV-2 virus, while cells that lacked ACE2 were not. In the micrographs, the ACE2 protein and the virus were stained two different colors to allow easy comparison and contrast. This comparison and contrast constitute the story, which constitutes the scientific evidence in question.

Storytelling doesn't necessarily need multiple micrographs. Snapshot micrographs can tell stories too, if the purpose of using them *is* to give a snapshot of

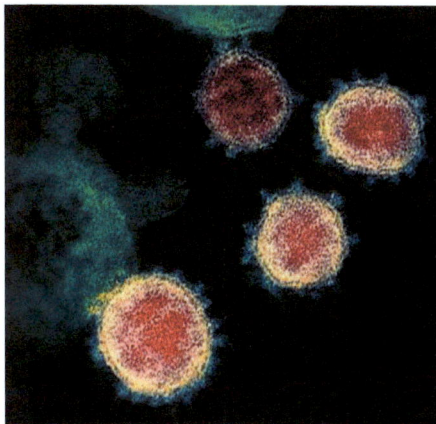

**Figure 3.13** SARS-CoV-2 snapshot micrograph as storytelling.

something and show its appearance and structure. In the early days of the COVID-19 outbreak, the National Institutes of Health wrote about the "crown-like" spike structure on the SARS-CoV-2 virus and the effort to develop vaccines that target the spike protein. Accompanying the article is the micrograph shown in Figure 3.13.

Although Figure 3.13 is a static snapshot, it provides the right evidence for its article: an outward appearance of the SARS-CoV-2 virus and its crown-like spikes that give the virus its name (another visual used in the article shows the detailed protein structure of the spike). In fact, one may say that in this case, the snapshot micrograph tells a story of "part and whole" – that is, here is the whole virus, and here is an important part that makes the virus what it is.

## Micrographs as Mysterious Wonder

As mentioned above, snapshot micrographs can be visually attractive. There are good reasons for that. Micrographs capture minuscule objects and microorganisms that we don't get to see in everyday life. And these objects and microorganisms often have unusual and intricate structures. In other words,

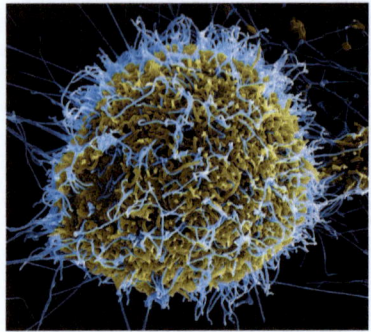

**Figure 3.14**    Micrograph of Ebola virus particles budding from a chronically infected cell.

the subjects of micrographs are often inherently interesting. Add to that the modern microscopic/imaging technologies, which are capable of producing highly magnified, generously colored, three-dimensional-looking micrographs, and the result is often stunning, a case of mysterious wonder unveiled.

Figure 3.14 is an example, showing Ebola virus particles in their unruly filaments, colored bright blue, budding out of an infected cell. Some of the virus filaments are curly and cling to the surface of the cell; others are wild and shoot straight out. The cell, looking sick with a yellowish-green color, is itself a curious object with a wiry texture, hopelessly smothered by a tangled web of viral filaments.

Similar – no, more mysteriously wonderful – micrographs are abundant on the internet. Winners of the Nikon Small World Competition, a global micrograph competition, often follow this theme: a mesmerizing web that is the vascular system of the mouse retina, curiously shaped veins and scales on a butterfly wing, clownfish embryos that look like amber trapped in translucent crystal balls. As one might expect, the winning criteria for the competition include, among other things, visual impact.

Then, there is micrograph labeled *as* art. Search for "micrograph art" online, and you can immediately start shopping for micrograph posters or canvas prints. Scientists, too, are well aware of the artistic and visual impact of their micrographs. Academic journals – *Nature Medicine*, *Science*, and *Cell*, to mention a few – feature mysterious and wonderful micrograph art on their covers.

What functions do these beautiful micrographs have? That is a complicated question to answer. If the micrographs are used in the right context to show, for example, the intricate structure of the Ebola virus, then they would serve not only as attractive visuals but attractive evidence. Life is mysterious and wonderful. The microscopic world surrounding us is no exception. Revealing that wonder to the public, who do not otherwise get to see it, seems not only appropriate but necessary. Mysteriously wonderful, these micrographs are well positioned to increase public interest in and engagement with life sciences.

At the same time, micrographs, I hope most will agree, aren't supposed to be abstract paintings. They have to make concrete sense, not only to dazzle a viewer. If we are so focused on making micrographs beautiful, we risk losing opportunities to make them informative. Consider this: In academic papers read by professional scientists, micrographs often come in small sizes in a tabular format to facilitate easy comparison. Frequently, texts, numbers, arrows, or other symbols are used to call out and explain elements of interest. Figure 3.15, from Troemel et al.'s study, which compares infected and uninfected roundworms (*C. elegans*) to identify a parasite known as *N. parisii*, shows such features. Micrographs that are designed to look beautiful can't have such features – they would ruin the aesthetics, even though nonspecialist readers, more so than professional scientists, need help reading micrographs.

Surrounded by mysteriously wonderful micrographs, we risk forming a false impression of the microscopic world. Beautiful micrographs are a product of conscious effort. One doesn't look into a microscope and automatically see beauty at every turn. Most microorganisms and cells are colorless and transparent. They appear in color because the samples are stained, or colors are added during post-processing. These colors help highlight microscopic details, but they are also very much chosen with an eye for artistic effect. Beautiful micrographs, then, are often stylized, just as Haeckel's drawings of radiolarians. If one is threatened by subjectivity, so is the other.

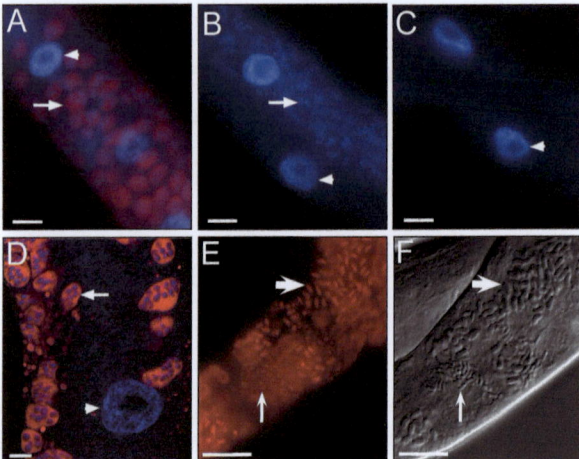

**Figure 3.15**  *N. parisii,* a novel microsporidia species, infects the roundworm *C. elegans.*

## Micrographs as Yuck

Besides mysterious wonder, there is another common trope used in popular science micrographs: yuck. Rather than aspiring to beauty, these micrographs inspire disgust and fear.

If this seems surprising at first thought, think again. Marketing and advertising industries, together with health campaigners, have long known that disgust and fear attract attention. A mouthful of nasty-looking germs sells toothpaste, a critter digging under toenails sells fungal treatment, a lung full of black tar (hopefully) dissuades people from smoking. Precisely because these images are unpleasant and fearsome, they invite (or rather compel) people to look.

Disgust and fear are easy to evoke through micrographs. What we can't see scares us, and we can't see the microscopic world. Plus, many minuscule agents *are*, by nature, not good for us: deadly viruses, disease-causing bacteria, blood-sucking insects.

These shady characters become downright repulsive when photographed through the scanning electron microscope. As mentioned earlier, these

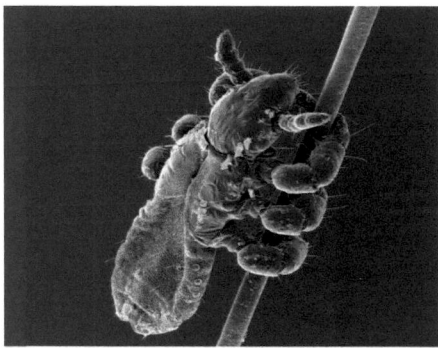

**Figure 3.16** Scanning electron micrograph of a head louse.

microscopes work by sweeping electrons across a sample and using the reflected electrons to generate an image of the sample's external structure. At low magnification they can capture an entire specimen and provide quite a bit of depth, making a micrograph look three-dimensional.

Figure 3.16 is an example. It shows a head louse clutching a human hair. The louse has a creepy-looking head like that of a venomous snake, pointy tentacles, robotic legs, irksome bristles, and a sunken stomach as if ready for a blood meal.

A comparison of Figure 3.16 and Figure 3.3, which are also micrographs of head lice, readily demonstrates their disparate intentions. In Figure 3.3, the lice are displayed as passive lab specimens, no props, no poses, no dramatic artificial color. The angles of the micrographs – a ventral (underside) view and a dorsal (back) view – provide unobstructed inspection of the entire bodies of the lice. The micrographs, then, are intended to function as a piece of the observational record, a way for people to study and identify head lice and their physical structures.

By contrast, in Figure 3.16 the head louse appears to be in action. Clutching a human hair, it looks intently ahead (at the scalp?!). It is artificially colored with an unpleasant grayish-purple. In this setup, the louse becomes an active predator, not a passive specimen. The side angle of the micrograph means that one can't see the louse's entire body. This micrograph, then, is intended to attract viewer attention more so than providing a visual record for study.

As long as we are making a comparison, let's note, quickly, the similarities and differences between Figure 3.16 and Figure 3.1, Robert Hooke's microscopic drawing of the louse. In both visuals, the louse is clutching a human hair. But in Hooke's drawing, the louse is belly-up, which renders it less threatening and more exposed for observation. This difference in angles, together with other stylistic differences in color and texture, reduces the yuck-index of Hooke's louse.

An online image search for "scanning electron micrograph" is guaranteed to yield many examples like Figure 3.16: a giant beetle head with bristled tentacles and compound eyes graces the homepage of the Santa Barbara Museum of Natural History; a bedbug with enormous legs and a gigantic mouthpart is featured on livescience.com; crawling mites and ticks with piercing-sucking mouthparts and plump bodies are shown on National Geographic; cancer cells are presented with gritty textures, sprawling spikes, and drab colors. If one heads over to stock image sites such as Alamy or Getty Images, one can find plenty more ready to order.

These micrographs make no pretense about being yucky. In fact, they seem pretty proud to be so. It is an intentional design choice to tap into our primal emotions.

Is that "okay"? Science communication is no toothpaste or fungal treatment, but it is a product no less. Whether it is science magazines or museums, they need to sell subscriptions, tickets, and advertising spaces. With the many leisure and entertainment opportunities available at consumers' fingertips, the public aren't exactly a "captive audience." Any means to direct their attention to science is valuable.

At the same time, Dorothy Nelkin's well-known critique – that media use exaggeration and sensationalism to "sell" science – looms. To put it crudely, modern science should not be a freak show. It is prudent for us to consider whether we are pushing ethical boundaries when micrographs function more as disgust-based sales strategies than informative or educational exhibits.

## Conclusion

As the little cousins of photographs, micrographs inherit some of the same misconceptions: that they are easy to understand and that what you see is the reality of the microscopic world.

In reality, micrographs are not at all easy for audiences who are not expertly trained in microscopy. Step one – making out the object of interest in a micrograph – can be difficult. The next step – making sense of the object – is no small feat either. It doesn't help when popular communication frequently uses snapshot micrographs, which zoom into a specimen but provide little context or clue in the way of cogent evidence.

Microscopic drawings, more so than micrographs, can provide interpreted evidence for easier understanding by nonspecialist viewers. Yes, these drawings can be stylized, but so can micrographs. From instrument choices to sample preparation to post-processing, micrographs are very much a product of human intention. In fact, today's popular science micrographs often function as symbolic artifacts, as pleasant or not-so-pleasant visual attractions. For this reason, micrographs are well positioned to create public interest in and engagement with science. For the same reason, we need to be aware, and at times wary, of micrographs' effect or the lack thereof.

# 4 Illustrations

Illustrations are a visual staple in life science communication. Despite being commonplace, they are in many ways a blackbox. They mask the creative – and scientific – decisions that go into making them. They present an end product that says, as it were, "this is how you look *through* life to its essence." The use of precise lines and explicit shapes helps to convey this scientific authority. In contemporary illustrations, pseudo-details such as colors and dimensions further prove that "this is what life *looks like*."

In reality, the best of illustrations can convey only *one* way of looking at life or, more precisely, one way of thinking about life. The worst of them, I'm afraid, convey preciously little constructive information at all. And in either case, the illustration contains more than "just facts." Individual choices, values, and styles, as we will see in this chapter, come together to determine looks.

## A Brief History of Illustrations

The term "illustration" is defined differently by different authors. The term easily overlaps with related terms such as "drawings," "schematics," and "diagrams." In this book, I use the term "illustrations" to refer to all hand-drawn and computer-generated visuals that employ elements such as lines, shapes, and colors to depict life science subjects, concepts, and processes.

Illustrations are the oldest type of life science visuals, because their creation requires no specialized instrument as in the case of photographs or micrographs. All it takes are pencils and skills. Some of the well-known early life science illustrations include Leonardo da Vinci's (1452–1519) studies of the fetus in the womb, which are two annotated sketches da Vinci created around 1511. Figure 4.1 shows one of them, depicting the human fetus inside a uterus. The uterus is cut open, showing the fetus inside. Da Vinci is considered the first person to have correctly depicted the position of the human fetus inside the womb.

Figure 4.1 reveals da Vinci's belief that life forms are comparable. The uterus pops open like a pea pot, and the smaller illustrations show the uterus being peeled open, like a flower or fruit, to reveal the seeds inside. Incidentally, the sketch mistakenly shows a cow placenta because da Vinci had dissected the cow uterus and assumed that the same structure existed in humans.

Shortly after da Vinci, another Renaissance genius, Andreas Vesalius (1514–1564), came along with his 1543 groundbreaking anatomy book *De Humani Corporis Fabrica Libri Septem* (*Seven Books on the Fabric of the Human Body*). Its famous muscle men we met in Chapter 1.

It is not coincidental that well-known early illustrations are often those of humans: As humans, we are naturally interested in our own physical structures and life forms. Indeed, these Renaissance masterpieces are far from being the earliest scientific illustrations of humans. On the other side of the world, in China, for example, illustrations of whole-body acupuncture points had existed at least 1,000 years earlier.

Chinese legendary physician Simiao Sun (541 or 581–682 CE; yes, legend has that he lived to over 100) created *San Ren Ming Tang Tu* (*Acupuncture Points in Frontal, Back, and Side Views*). Figure 4.2 shows the frontal view. Each tiny dot and accompanying label marks an acupuncture point. The lines that connect the points are major nerves, which were color coded in the original illustration. Together, Sun illustrated 650 acupuncture points (282 in the frontal view, 194 in the back view, and 174 in the side view), which represent 349 unique acupuncture points.

In addition to humans, plant and animal anatomies are also favored topics of early life science illustrations. German-born nature illustrator Maria Sibylla Merian (1647–1717) is known for her work on insects and plants. In her book *Der Raupen wunderbare Verwandelung, und sonderbare Blumen-nahrung* (*Caterpillars, Their Wondrous Transformation and Peculiar Nourishment*

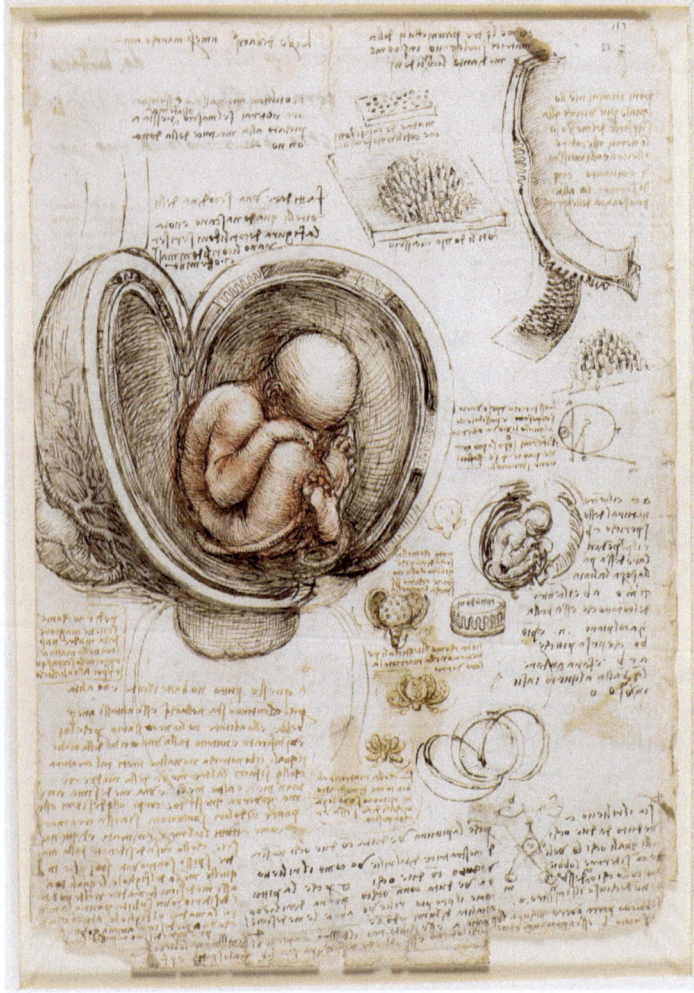

**Figure 4.1**   The fetus in the womb by Leonardo da Vinci.

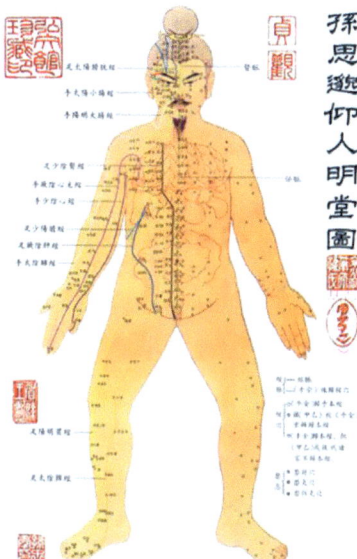

**Figure 4.2**    Simiao Sun's acupuncture points in frontal view (reproduction).

*from Flowers*; first volume published in 1679 and second volume in 1683), Merian depicted the metamorphosis of moths and butterflies, along with their plant food sources (Figure 4.3). Her detailed and precise illustrations of metamorphosis contributed to the development of modern entomology.

Merian's work – and da Vinci's and Vesalius' work too – are what one may call "naturalistic." That is, they pay attention to observational details and try to depict subjects as they appear, making them "life-like."

Compared with these examples, modern life science illustrations often look more streamlined and purposefully instructional. Figure 4.4, for example, shows the basic structures of the brain in a cross section. Rather than trying to depict the brain as it would look on a dissection table, with realistic colors and texture, the illustration provides a skeletal outline and focuses on demarcating the different structures. This style reflects the modern ethos of interpreted visuals, where looking real is deemed less informative than highlighting

**Figure 4.3**    Moth and mulberry tree, by Maria Sibylla Merian.

essential features. Interestingly, Sun's acupuncture point illustration, despite being over 1,000 years old, sits somewhat in between the naturalistic Renaissance illustrations and streamlined contemporary illustrations. Sun used simple outlines to depict the human body; at the same time, he captured realistic details such as the man's hairstyle and facial expression.

Not all contemporary life science illustrations look as streamlined as Figure 4.4. There is another branch of illustration that, although a result of interpretation, looks very much the opposite. Figure 4.5 is an example, showing the structures of an animal cell. Figure 4.5 doesn't try to imitate a grainy micrograph to show what a cell looks like under the microscope, nor is it

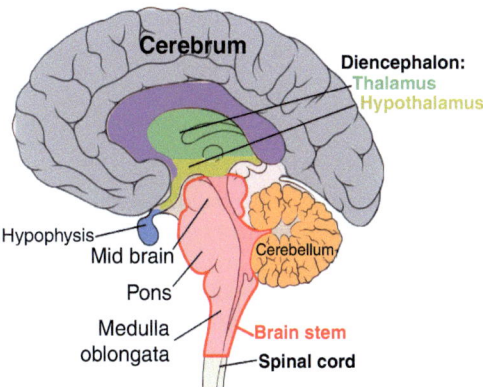

**Figure 4.4**    Basic structures of the brain.

**Figure 4.5**    The anatomical structure of an animal cell.

content with outlining the skeletal appearance of the cell. Instead, we have colorful, exquisite, and ornamental structures that look very fanciful and immediately draw the reader's attention. Illustrations like Figure 4.5 represent contemporary scientists' and science illustrators' conscious efforts to interest and engage nonspecialist readers in science.

In the rest of this chapter we will revisit these different styles of life science illustration to examine their effect and question our assumptions about them.

## Illustration as Interpretation

Compared with photographs and micrographs, illustrations are the quintessential "interpreted visuals." They represent scientists' interpretations of what their eyes perceive or what their minds conceptualize. Emerging from this active interpretation, illustrations bring to the foreground what is considered a subject's most essential visual elements and disregard distracting details. As a result, the idea goes, the depicted subject will be easier for nonspecialist viewers to understand.

With Figure 4.4, for example, if we use a photograph of a dissected brain, the messy and complex organic brain matter will make it harder for readers to separate the different structures for easy viewing and understanding. This issue is compounded by technical limitations. Despite ongoing technological improvements, cameras are, in many ways, no comparison to the human eye in terms of flexibility and versatility. Any imperfections in lighting, specimen position, or distance can detract from the quality of anatomy photographs. By contrast, illustrations are more forgiving because they are the result of interpretation. Shadows, for example, can be easily omitted to bring out the essential parts of an object.

It is important to note that what counts as "essential" visual elements for easy understanding is not a given. We examine this point in detail in the next section. For now, it is useful to add that cognitive interpretation is not all that illustrations can do. In addition to filtering out distracting details to help with cognitive understanding, illustrations can also filter out details to help with emotional reception.

Anatomy provides a case in point. Anatomy photographs, especially those of the human body, are often raw and graphic. They remind us of our corporal existence – and one day, the inevitable end of that existence. They remind us of the vulnerability of our body, the way it is easily corrupted by diseases and trauma. These reminders can be particularly problematic in popular science communication, where the audience range in age, life experience, and emotional readiness.

Illustrations, with their clean lines and shapes, can filter out the blood and flesh – and the unpleasant emotions along with them. So it is that we can stare at Figure 4.4 with little, if any, adverse emotional reaction. Even if the illustration is about a diseased brain – with a tumor growing inside it, for example – the visceral reaction it creates would still be minimal.

## Simplified Icons Mask Individual Interpretation

Icons are simplified, pictorial representations of reality. They are abundant in everyday visual signaling. An icon of an airplane by the highway, for example, signals the exit for an airport. Inside the airport, icons of luggage, taxi, escalator, etc. designate different services and facilities.

Except for the most naturalistic depictions, scientific illustrations rely on icons. Indeed, this is where the instructional value of illustrations comes from. By reducing complex visual realities to their essential features, icons are generally acknowledged to enhance comprehension and information recall. This belief can be traced to neurological studies of working memories, which demonstrate that our working memories have a limited capacity. Visuals with complex surface details can overload our brains and become barriers to comprehension. Drilled down to bare essentials, simple icons are supposed to be easier to visually process and comprehend.

Figure 4.4, for example, uses simple lines, curves, and shapes to depict a complicated organ. A small blob signals the hypophysis, a simple oval represents the thalamus, and a "wrinkled" circle/ring is the cerebellum. Distinct colors separate the different structures. No obvious shadows or dimensions were attempted. Everything is laid out in an efficient – and, as mentioned above, an unemotionally efficient – manner.

Compared to naturalistic drawings, such as the moths and mulberry tree depicted in Figure 4.3, illustrations like Figure 4.4 *look* more scientific, more authoritative. By reducing the physical reality to abstract lines and shapes, they say, to borrow Gross and Harmon's words, that what is depicted is "not a means of looking *at* the world, but a means of looking *through* it to its causal

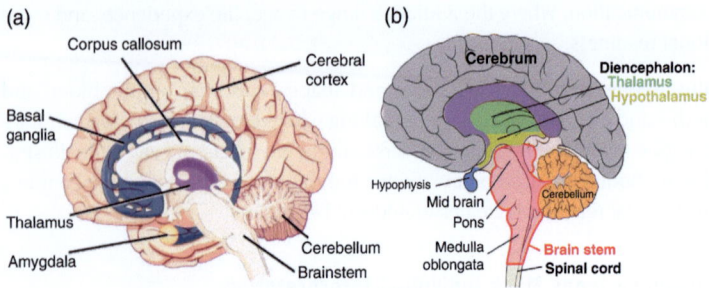

**Figure 4.6**  Two interpretations of the basic structures of the brain.

structure" (p. 81). In other words, these illustrations represent the ethos of modern science as the means to dissect nature and life, to reveal the truth under the surface. Interestingly, this visual impression effectively masks the nature of illustration as individual interpretation, as only one version of reality.

Take the example of acupuncture points. Traditional Chinese medicine is equivocal on the total number of acupuncture points in a human body – let alone what or where they are. Some say 300+ points, others say 700+ points. Yet, a given illustration (any given illustration), by committing dots to page, eradicates these differences and presents a finalized conclusion: Where a dot is, an acupuncture point must be.

Even when a topic is not ostensibly under debate, like the structure of the brain, a given illustration is still a particular interpretation. Figure 4.4 illustrates one perspective of what counts as basic brain structures. Another illustration of the same topic may well have different opinions on what is "basic" and worth illustrating (see Figure 4.6(a), compared to the original Figure 4.4, shown here as Figure 4.6(b)).

## Simplified Icons vs. Concrete Icons

The icons used in Figure 4.4 are not the most simplified there are. Going back a few decades, it is common to see illustrations like Figure 4.7 in popular science magazines, illustrations that really push the envelope of iconic

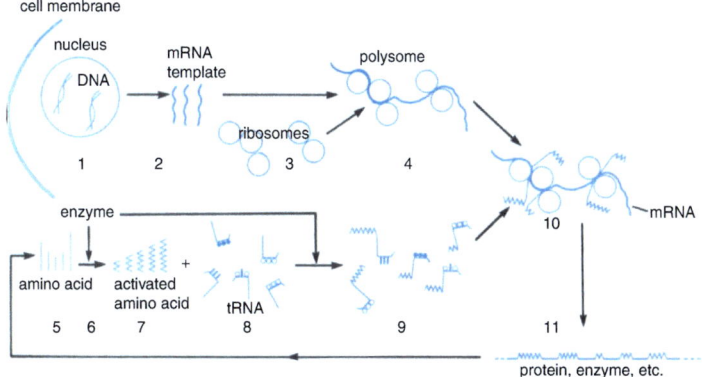

**Figure 4.7**    A 1970s minimalist depiction of protein synthesis.

efficiency. Originally published in a 1975 *American Scientist* article, Figure 4.7 depicts the process of protein synthesis, a complex, 11-step process of how DNA directs the production of proteins.

Based on its drawing style, Figure 4.7 is what one may call minimalist. Gone are any visual cues of organic subjects (such as soft textures or irregular shapes). What's left are straight lines, perfect circles, regular zig-zags, and, at most, curvy lines. This is, if you will, the "stick-figure" version of life science illustrations.

If reducing visual complexity and removing distracting details will enhance visual processing and comprehension, then Figure 4.7 ought to be the easiest to understand among all the illustrations we have seen. This, I think most readers will agree, is not the case. Part of this is because Figure 4.7 depicts a dynamic process, whereas the other examples focus on static objects (Figures 4.1 and 4.3 arguably nod toward a general process). To understand an object illustration, we generally only need to figure out appearances and structures. By contrast, to understand a process illustration we need to figure out what comes first, next, and last; which agents are interacting with which other agents; how they interact; and what the outcomes are. It certainly doesn't help that the process depicted in Figure 4.7, protein synthesis, is a complex one to start with.

Aside from this inherent difference, from a visual perspective, Figure 4.7 also *feels* like it would be difficult to understand. This is because the cognitive benefits of simple icons are overshadowed by the icons' lack of concreteness. Concreteness, as McDougall et al. explain, refers to the extent that icons represent things that viewers are familiar with in the physical world: plants, animals, consumer products. When icons are concrete, viewers can apply what they already know about the "thing" in real life to understand the intended meaning of the icon. The human brain is, relatively speaking, familiar to us at least in its overall shape housed inside the skull. It is also conceptually familiar as the organ that governs our cognition and emotion. This familiarity can be seen by the extent that "brain" has entered our daily conversation (we say that "my brain is fried" or someone "has the brains"). Because of this, the brain structure illustration of Figure 4.4 has relatively higher icon concreteness. For the same reason, the acupuncture point illustration of Figure 4.2 has even higher icon concreteness.

This, however, is not true with the various agents involved in protein synthesis. Whether it is DNA, mRNA, tRNA, or ribosomes, they are not visible or commonly referenced in everyday life. As such, readers will not be able to easily connect the icons in Figure 4.7 to their physical reality and intuit the underlying meanings. In fact, a significant amount of background knowledge is required to follow the icons: For example, how does the double-line DNA become single-line mRNA, or how does tRNA seem to "pick up" amino acids?

Last but not least, because Figure 4.7 uses austere lines, shapes, and arrows all in one color (aside from the color black), it *feels* clinical and opaque. In other words, regardless of its actual content, it *feels like* it would be difficult to understand and would require serious study. This visual impression is aligned with the prevailing attitude toward public understanding of science in the 1970s and 1980s, the so-called deficit view (see Chapter 1). According to this view, the public lacks essential knowledge of science and must diligently catch up, and scientists are the experts and must fulfill their responsibility to educate. Figure 4.7, accordingly, unapologetically lays out the details of protein synthesis and challenges readers to study and understand it.

In retrospect, it is easy to see the problem with Figure 4.7 and illustrations like it. As Jen Christiansen, graphics editor at *Scientific American*, put it, they are "off-putting to a non-specialist audience" (p. 49). Scientific visuals that look utterly unfamiliar, inaccessible, and irrelevant to everyday people have a slim chance of inspiring study and understanding.

## Pseudo-concreteness

Guided by the need to interest and engage the public, contemporary popular science illustrations have, by and large, shed the austere, minimalist look. Instead, vibrant colors, three-dimensional perspectives, intricate textures, and dramatic composition have become the norm – as shown in Figure 4.5, the structure of an animal cell. Compared with Figure 4.4, the structure of the brain, Figure 4.5's visual complexity is more than doubled. Instead of adopting a flat outline, the cell looks like a squeeze ball cracked open, with a wedge taken out to reveal its inner structures. The inside of the cell is filled with blue liquid, and floating on top of it are various curious-looking structures. The endoplasmic reticulum looks like folded cake icing; the centriole looks like intricate gear one would find inside sophisticated machinery; the mitochondria look like curiously shaped food trays from a cute doll house. Everything has a pleasant, smooth texture, and lighting is expertly used to create glows and shadows.

The same visual style can be used to illustrate complex processes such as protein synthesis, the topic of the austere, minimalist Figure 4.7. Look, for example, at Figure 4.8. This illustration sports a gradient of dark colors as its background, evoking an outer-space look. This is a common visual trope in contemporary life science illustrations, and one that is purposefully chosen. Space is humans' next frontier, and it symbolizes tremendous efforts with exciting possibilities. By aligning with that context, life science illustrations suggest that life science research, too, is the next frontier, tremendous and exciting. Against this space-like background, giant ribbons of DNA and mRNA swirl around with colorful columns for nucleotide bases, tRNA are lined up as intricate jigsaw puzzle pieces, and amino acids are strung together as solid marbles.

One final example of this elaborate visual style, an illustration that has saturated mass media in the year 2020, is the SARS-CoV-2 virus illustration designed by the US Centers for Disease Control and Prevention (CDC) (Figure 4.9).

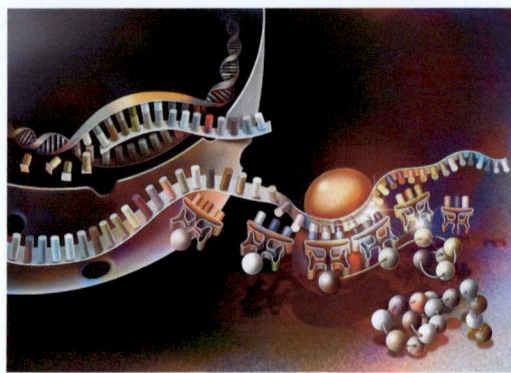

**Figure 4.8**    Contemporary depiction of protein synthesis with pseudo-details.

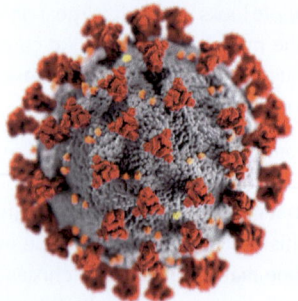

**Figure 4.9**    A beauty shot of the severe acute respiratory syndrome coronavirus 2 (SARS-CoV-2).

Figure 4.9 is what designers at CDC call a "beauty shot": a detailed close-up of a single viral particle. The designers told *New York Times* writer Cara Giaimo that they chose a stony texture for the virus so that it looked like "something that you could actually touch." Gray was used for the virus' fatty membrane, and several bright accent colors were used to depict viral proteins to make them stand out. The spike protein (or S protein), which looks like a crown and allows the virus to attach to human cells, is colored a deep red. Lighting was calibrated so the spikes cast long shadows to "help display the gravity of the situation and to draw attention." The membrane protein (or M protein), which is embedded in the membrane and gives the virus its shape, is colored orange. Finally, the envelope protein (or E protein), which plays multiple roles, including helping the virus to assemble in the host cells, is colored yellow.

From Figures 4.5 to 4.8 to 4.9, the subjects being depicted, whether the endoplasmic reticulum or spike protein, are not familiar to everyday readers. Therefore, the icons used to represent them will, by definition, be low in concreteness. Yet that is not the *visual impression* one gets when looking at the illustrations. On the contrary, the icons look very concrete, like something one could touch, to borrow the words of the CDC designers.

This effect is achieved by what I call pseudo-details: textures, shapes, dimensions, colors, shadows, compositions, all of those visual details we saw above. These details are "pseudo" because they are artistic additions, not something someone has observed. They are not exactly "false" details though because, often, they have certain scientific bases. What do I mean by this? Life in the microcosm defies common ways of "seeing." The appearances of molecules are deduced based on atomic structures. The structure of the coronavirus spike protein, for example, is shown in Figure 4.10.

Figure 4.10 is what's known as a ribbon diagram, which shows the overall structure of a protein backbone in three dimensions. The different shapes in the diagram represent different chemical structures. A coiled ribbon, for example, represents an alpha helix, which is a common structure in proteins and consists of a single chain of amino acids. The chain is coiled, like a spring, with each "loop" in the coil stabilized by hydrogen bonds.

Appearance-wise, Figure 4.10 looks very different from the stony-textured red spikes in Figure 4.9. The two do share some resemblance in their overall shape, but without artistic elaboration one could not have bridged the two.

By this, I do not mean that Figure 4.10 is the *real* spike protein. Ribbon diagrams are only one way to visualize a protein. With changed settings in modeling software, we can change a protein's appearance. For example, we can add a surface cover to the structure – imagine draping a light, silky cover

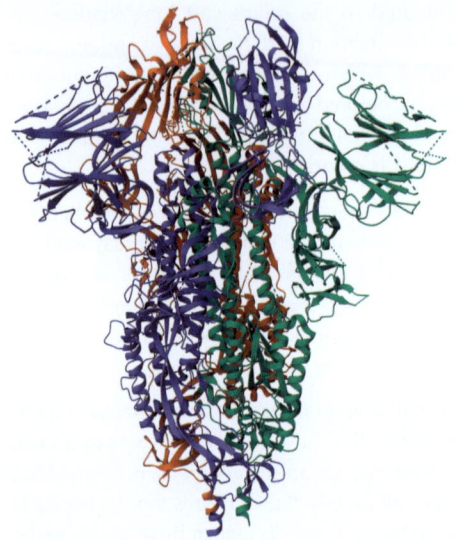

**Figure 4.10**    A ribbon diagram of the coronavirus spike protein.

over Figure 4.10 and letting it hang to bring out all the contours in the structure. The resulting look would be very different.

Pseudo-concreteness makes contemporary life science illustrations appear more relatable to our everyday lives and thus more accessible. As such, they provide public readers with an entry point and stand a better chance of engaging readers. The success of the CDC's coronavirus illustration all but proves this point. During the COVID-19 pandemic, this illustration has spread far and wide around the world to raise public awareness about the pandemic. Creative folks even turned the viral pattern into cookies, knitting projects, and headwear accessories.

At the same time, pseudo-concreteness adds to the misconception that contemporary life science illustrations depict the reality of nature and life. The vivid shape, touchable texture, and distinct color all give a convincing visual impression that this *is* how things look, as opposed to being someone's interpretation of scientific findings. From a technological standpoint, this seems to make perfect sense: The illustrations look so polished that one assumes they are the product of sophisticated new technologies, technologies that allow us to see what we couldn't see before.

Still, so what? So public readers take the realism of life science illustrations for granted – what's the big deal? They are more likely to be interested and engaged in the sciences, aren't they? That may very well be true, but potential issues can arise when pseudo-details become the so-called "seductive details."

## Seductive Details and Risks of Miscommunication

Seductive details are highly interesting details that are only tangentially relevant to the purpose of a communication. Harp and Mayer gave the example of telling an anecdote of being struck by lightning in a text that explains the formation of lightning. Some scholars argue that seductive details can distract readers, disrupt information processing, or activate inappropriate prior knowledge schemes to mislead readers.

The pseudo-details that we see in contemporary illustrations run the risk of becoming such seductive details. A case in point is Figure 4.11.

Figure 4.11 appears in a National Institutes of Health news article about the Fragile X syndrome and associated genetic disorders. These disorders are caused

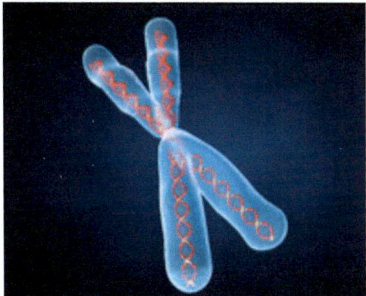

**Figure 4.11**    Misleading illustration of a chromosome.

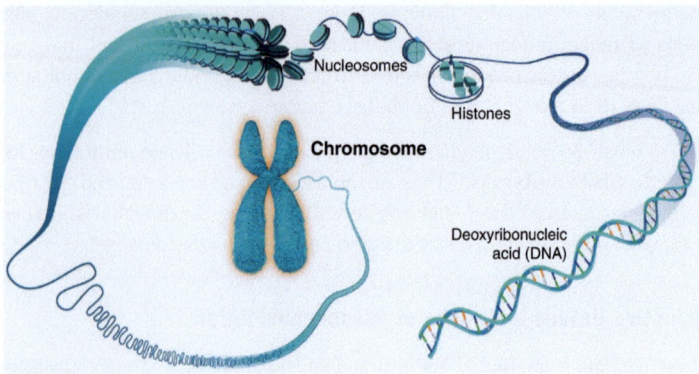

**Figure 4.12**  Structure of a chromosome.

by mutations in a gene on the X chromosome and lead to various symptoms, including intellectual and developmental disabilities. Accompanying the article, Figure 4.11 supposedly illustrates the X chromosome, or at least a generic human chromosome.

As is typical, Figure 4.11 uses a black/blue, space-like background. DNA is depicted in bright red double helixes. Surrounding the helixes are transparent, glowing, tube-like structures, which presumably denote chromosomes. While a geneticist will "get" this illustration, for the general public, this attempt to give chromosomes a tangible, interesting shape will likely activate inappropriate prior knowledge. That is, it readily cues a reader that chromosomes "house" DNA or that DNA stays inside hollow chromosomes. In reality, chromosomes *are* densely packed DNA and proteins. As shown in Figure 4.12, each chromosome is made of a long DNA molecule and proteins known as histones. Eight histones come together to form a tiny spool-like structure, and DNA winds tightly around it – this basic unit is known as a nucleosome. Without this tight packaging, DNA would be too long to fit into cells.

If we want to be critical (which we probably should with high-impact communications), even the CDC's popular coronavirus illustration risks

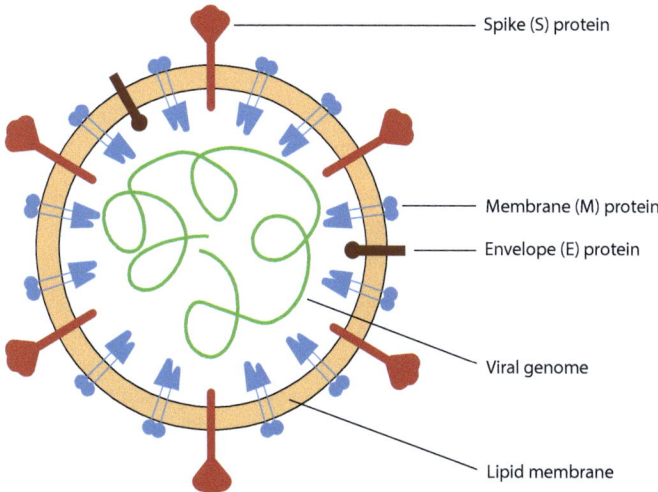

**Figure 4.13**    Illustration of coronavirus without pseudo-details.

misleading readers. Looking at Figure 4.9, one is struck by the fact that the viral particle has many spike proteins on its surface. This would seem to make sense, since the crown-like spike is what gives the virus its name. But, actually, the membrane protein (the small orange dots in Figure 4.9) is the most abundant in the coronavirus, as Figure 4.13 attempts to illustrate.

The designers at the CDC know very well the relative abundance of the viral proteins, but decided to foreground the spike protein. As they told the *New York Times* reporter, this is because the spike protein is presumably responsible for the virus' rapid spread. It doesn't hurt that its long shadows "help display the gravity of the situation and . . . draw attention." As one can imagine, the small membrane protein wouldn't quite create such a visual impact.

Figure 4.13, by contrast, is more interested in conveying information than drawing attention. It used simpler icons to show the relative frequencies of different viral proteins and callout texts to name different viral components.

Despite its reduced number, the spike protein *still* looks significant with its larger size. In other words, I wonder if the CDC could have struck a better balance between visual attraction and structural details in this case.

## Seductive Details and Visual Metaphors

Seductive details that (inadvertently) mislead or miscommunicate are, by definition, still trying to convey scientific information; seductive details that have gone metaphorical, it seems, don't even try.

Metaphors are a common figure of speech. They use a familiar concept or a concrete concept to explain an unfamiliar or vague one. For example, we may say that someone is a "diamond in the rough." The expression makes sense because of the underlying similarities between the two entities: a person who has a lot of potential but needs further refinement is like a rough diamond in need of cutting to reveal its full brilliance. Because concepts like "potential" and "refinement" are relatively vague and hard to pinpoint, we replace them with the concrete and visible action of cutting a diamond. The familiar or concrete concept that is used to explain another concept is known as the source of the metaphor. The unfamiliar or vague concept that needs explanation is known as the target. So, when we say someone is a diamond in the rough, the diamond is the source, and the person is the target.

As Andrew Reynolds writes in *Understanding Metaphors in the Life Sciences*, the disciplines of the life sciences are no stranger to metaphors. Scientists frequently use analogical thinking to develop theories, describe observations, or explain nature and its mechanisms. For example, "cells" are called cells because when Robert Hooke, the creator of the giant flea and louse (see Chapters 1 and 3), examined a piece of cork under his microscope, he saw many little pores or boxes with thin walls. Hooke thought these boxes look like tiny monastery rooms, or cells, occupied by monks and decided to call them cells. Today, the metaphor has become so pervasive that we take the name for granted and stop seeing it as a figure of speech. Or, as linguists would say, it has become a dead metaphor.

A more recent life science metaphor, not quite dead but very common, compares DNA to life's blueprint. The idea is that DNA provides the biological information to build proteins, which go on to enable life forms and functions – much the same way we use blueprints to construct buildings.

Among the scientific community, metaphors like these are a bit of a controversial topic. On one hand, they provide entry points for scientific discovery, comprehension, and explanation. As Taylor and Dewsbury argue, metaphors can bridge phenomena in the macrocosm (e.g., the biosphere) or the microcosm (e.g., molecules) with phenomena in our day-to-day world, which makes metaphors especially valuable in conducting and communicating science. On the other hand, metaphors, by likening two different things, are inherently imperfect. The DNA that we were born with doesn't provide all the information for life. The external environment, for example, can affect the workings of DNA by turning genes on or off. Because of this inherent imperfection, metaphors like the "DNA blueprint" risk creating fundamentally flawed explanations.

While scientists are well aware of this conundrum posed by verbal metaphors, not much has been said about visual metaphors in science. Visual metaphors work in similar ways to verbal metaphors. That is, a familiar or concrete image (the source) is used to explain an unfamiliar or vague one (the target). The advertising industry has long used visual metaphors to sell products. Silky fabric, for example, is used to sell lotions, suggesting the smooth and luxurious nature of cosmetic products. The beverage industry, in particular, is fond of visual metaphors. Snow-covered mountains sell refreshing beer; a pair of wings sells explosive energy drinks; basketball hoops sell exciting and fun soda.

What do visual metaphors look like when used in the life sciences? Figure 4.14 is an example, illustrating the concept of gene editing. Gene editing attempts to alter our genes, or functional segments of DNA, for therapeutic effect (or, as the fear goes, for nontherapeutic doping). A mutated gene may be replaced with a healthy copy, a disease-causing gene may be inactivated, or a new gene may be introduced to help treat a disease.

A recent development in gene editing technology is CRISPR-Cas9. Put simply, CRISPR-Cas9 is a DNA-cutting mechanism that naturally exists in bacteria as a defense mechanism against viruses. CRISPR (which stands for clustered regularly interspaced short palindromic repeats) are short, repetitive DNA

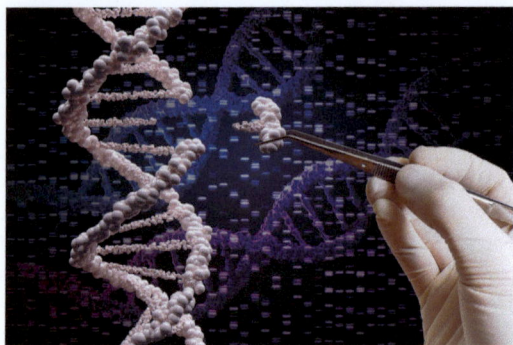

**Figure 4.14**  A gene editing visual metaphor.

sequences sandwiched between so-called "spacer" sequences. These short repeats match viral DNA so they can guide Cas9, an enzyme, to cut the DNA of invading viruses and disable them. Inspired by this natural mechanism, scientists combined Cas9 and a matching sequence called the guide RNA. Using the guide, Cas9 can target specific DNA, cut it, or help change it.

Figure 4.14 illustrates these ideas metaphorically. In the foreground is a strand of double-helix DNA in pearly white – each atom looks like a shining pearl. A gloved hand (presumably that of a scientist) holding a pair of tweezers reaches in and snips a short strand of DNA from the double helix. In the background, more double helixes and what looks like electrophoresis bands (electrophoresis is a laboratory technique used to separate DNA, RNA, and protein molecules of different sizes) float on a black/blue background, evoking the common outer-space theme.

In Figure 4.14 the source image is the hand using tweezers to pick up something small; the target image is editing or changing our genes. The former is easy, familiar, and concrete; the latter is complex, unfamiliar, and vague. Hence, the metaphor.

Surely, everyone *knows* that this is a metaphor? I don't know *how* sure we can be. National and international studies suggest that the public tends to understand DNA and genetics via the general concept of inheritance, but is not familiar with finer details. For example, according to a 2017 BBVA Foundation study, only 44.9% of US adults self-identified as having a complete understanding of DNA. People in the European Union fared somewhat better at 52.8%. In Middleton et al.'s 2020 study of 36,268 individuals from 22 countries across the world, the majority of participants (anywhere between 54% and 88%) in 20 countries said they were unfamiliar with concepts of DNA, genetics, and genomics. Italy and the US were the only countries where the minority (42% for both countries) said they were unfamiliar with these concepts.

If readers do not know, for example, where DNA is, how small it is, and how it can be accessed, then Figure 4.14 risks being taken literally. That is, we can physically manipulate DNA precisely and easily, like picking something up with a pair of tweezers. Even barring a literal interpretation, Figure 4.14 still advertises a false sense of security and certainty about gene editing. If gene editing is as easy as bringing a pair of tweezers to DNA, then we will have no problem applying it to treat or prevent diseases. By extension, there should be no problem applying it to "better" ourselves, making us smarter and stronger, for example.

But things *aren't* this easy. Gene editing is full of complications. For example, it can land on the wrong DNA sequence and cut something off-target, causing unintended mutations or deletions. Or cutting a DNA sequence may result in cell death rather than an intended DNA change. Immediate adverse effects from gene therapy include death, and delayed adverse effects have been reported years after treatments.

Some readers may think that I'm being overly critical here. So Figure 4.14 doesn't convey factual information, can we not use it as a fun, intriguing visual decoration that "spices up" science communication? Yes, we can, but there may be a price to pay. As linguists George Lakoff and Mark Johnson tell us, metaphors are powerfully effective at regulating our underlying conceptual systems. When we are used to thinking of time as money, we develop a certain value system about time that we live our lives by. Similarly, if we routinely see gene editing illustrated as a simple action, we will believe that gene therapy (and medicine and science in general) is simple, familiar, and *not* worth asking specific questions.

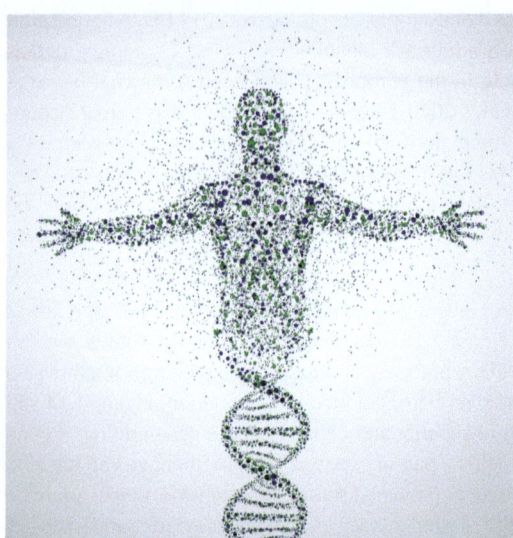

**Figure 4.15** A DNA–human body visual metaphor.

In addition, I am concerned that the "fun" nature of visual metaphors is precisely a problem. Consider Figure 4.15, which is a common type of visual metaphor where DNA is intertwined with life forms.

What exactly does this illustration mean? Clearly, this is a visual metaphor because we know that human bodies do *not* look like a double helix. If we go by the definition of metaphors where the concrete image is the source and the vague image is the target, then the human body ought to be the source and the DNA strand should be the target. If so, fusing the two images would seem to mean that "DNA is like the human body" or "DNA is connected to the human body."

Such statements, if given verbally, would sound odd and vague, if not totally meaningless. Yet, presented as a visual metaphor, they suddenly appear intriguing. This is because readers have to exert mental energy to decipher the visual metaphor and that process helps to elicit positive reactions.

Advertisers know this well. As marketing scholars Barbara Phillips and Edward McQuarrie explain, consumers respond with pleasure to images that artfully deviate from expectations, like a can of energy drink growing wings.

When readers are satisfied with the process of solving a visual puzzle, they are less likely to ask specific questions, such as how exactly DNA is related to human bodies (or how exactly a brand of energy drink works). And not asking such questions contributes to misunderstandings. If life forms are, literally, made up of DNA strands, then we have nowhere to go but accept genetic determinism and believe that our lives are determined by nothing but our genetic materials. Yet, as mentioned above, DNA doesn't wield this singular power.

If visual metaphors don't work, what alternatives do we have for illustrating complex topics? We can't go back to the 1970s minimalist illustrations, can we? No, we can't. The trick, science illustrators say, is to balance visual attraction with information richness. For example, *Scientific American*'s Christiansen suggests using visual metaphors *in combination* with technical visuals, such as using pulled rubber bands in combination with graphs to illustrate the relationship between velocity and distance. The visual metaphor provides readers with a welcoming entry point, upon which they are more likely to peruse the technical details.

Or, how about using illustrations that are informative but also *well designed* to be easily understandable? "Interest" is not a mere emotional arousal toward things that look fun. Instead, or in addition, we can interest readers by helping them to understand information and realize the significance of that information. This is what some scholars call "cognitive interest."

Consider Figure 4.16, which illustrates how CRISPR-Cas9 works. It is relatively fun to look at, with its use of a decorative double-helix DNA strand, bright colors, and an eye-catching composition (thanks to the swirling DNA). From a cognitive interest standpoint, simple icons, numbered steps, and short text notes all help to illustrate the basic steps of gene editing. Yes, as Figure 4.16 shows, gene editing is a complex technology, but not *that* complex that we *must* hide it behind a metaphor. Even if Figure 4.16 leaves readers with unanswered questions, that is a far better outcome than vague satisfaction. The former promises true interest and engagement; the latter risks complacency and misunderstanding.

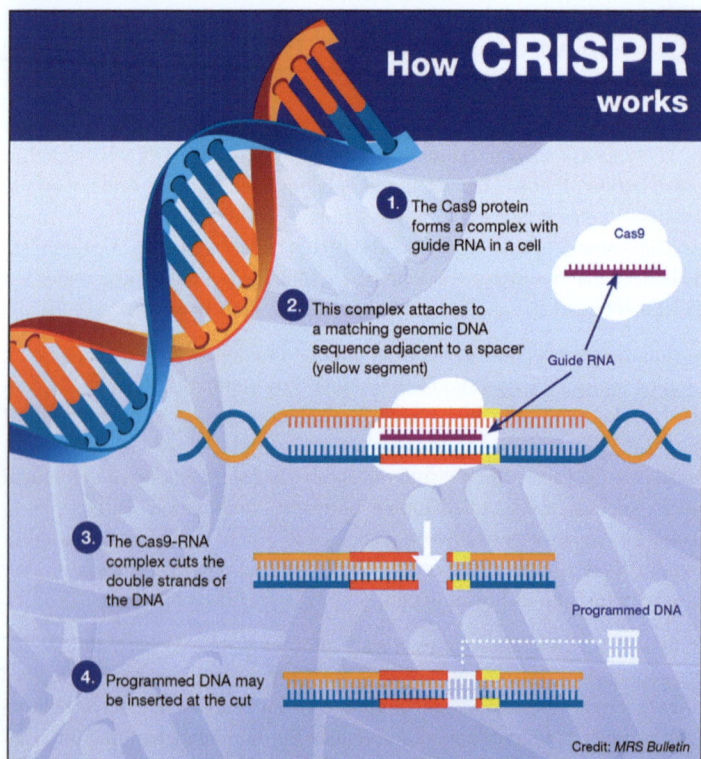

**Figure 4.16** How CRISPR works.

## Conclusion

Illustrations have a history as long as that of science itself, and their significance in the life sciences will only continue. Because illustrations "peel open" the exterior appearance of nature and life, they are charged with explaining complex structures and processes. That charge comes with scientific interpretation and stylistic choices, including what to illustrate and how to do so. Illustrations, then, are always going to be a version of reality, no matter how scientifically authoritative they are or appear.

The charge to explain the complex inner workings of science can make some illustrations difficult to follow for nonspecialist readers. In modern and contemporary life science communication, we have seen the use of simple icons, pseudo-concreteness, and seductive details to make illustrations more accessible and attractive. These stylistic choices have distinct effects that we will do well to be aware of. Ultimately, the primary purpose of science illustrations is to convey scientific information, and visual attraction must be balanced with that purpose.

# 5 Graphs

Graphs – such as line graphs or bar graphs – convey numerical data. They are commonly used in life science communication as well as other communication contexts, such as when conveying stock market data, crime statistics, or real estate trends. The prevalence of these graphs doesn't mean, as some may assume, that they are always easy to understand. Depending on design choices, some graphs will be able to shed light on important numerical data for public understanding of science, while others are likely to confuse or leave readers with a heightened conviction that science is an inaccessible enterprise.

In addition, just because graphs present "hard" numbers does not mean that they will always tell the same stories. Depending on design choices, graphs can create very different visual impressions – and therefore cultivate different beliefs – with the same set of numbers.

Visual "economy" is also a matter of judgment when it comes to graphs. It is commonly believed that graphs should be economical, that visual decorations such as the use of pictorial elements in pictographs are juvenile, and that graphs should convey the most ideas in the least amount of space. While sound in principle, these beliefs can also lead to stereotypes.

This chapter overviews the most common types of graphs for popular life science communication and discusses issues (as well as good intentions) that can complicate nonspecialist viewers' understanding of graphs. It also discusses pictographs and resists the stereotype that these graphs are unsuitable for serious science communication.

I want to give advance warning that the graphs used in this chapter tend to be about morbid topics such as deaths and diseases. This is because these data are widely available and frequently the topic of graphs that exist in the public domain.

## Commonly Used Graphs

Before we look at commonly used graphs in popular life science communication, it is useful to note that numerical data can also be organized into tables. In fact, if we use software such as Excel to create graphs, we would first enter the data into a tabulated format before the software can turn them into graphs.

This tabulated data are often not displayed in popular science communication. While tables have the ability to present diverse (numbers as well as text), precise, and large amounts of data, they are not good at showing trends or patterns. Only when the amount of data is small and their relationship straightforward can tables effectively demonstrate numerical trends. But such is often not the case with life science data. Indeed, if the data set is small and its pattern simple, we can simply state that information in text and save the real estate that a table must take up on the page.

In addition, tables are less eye-catching and attractive. There are not a lot of opportunities to use colors, for example, and the general format of a table (rows and columns) lacks visual novelty. This further deters the use of tables in popular science communication.

Now, the graphs. The three most commonly used graphs in popular science as well as everyday communication are line graphs, bar graphs, and pie graphs. The invention of all three is generally credited to the Scottish engineer and political economist William Playfair (1759–1823). In his 1786 and 1801 publications, *The Commercial and Political Atlas* and *Statistical Breviary*, respectively, Playfair used these graphs to convey a range of economic and demographic data: population, revenue, imports, and exports. Playfair heartily agreed that graphs are better conveyers of data than are tables. As he put it, "making an appeal to the eye when proportion and magnitude are concerned, is the best and readiest method of conveying a distinct idea" (p. 4).

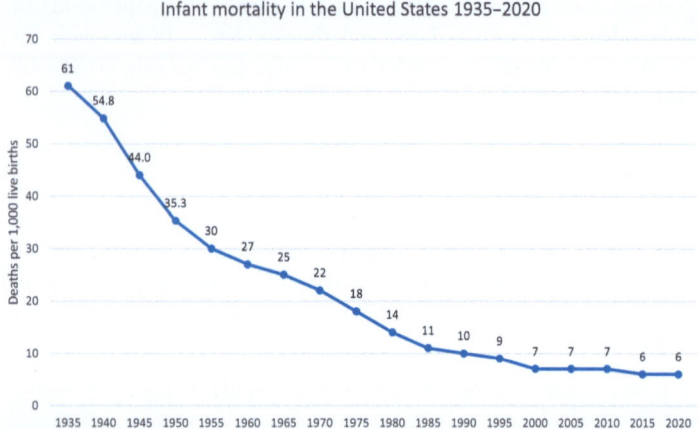

**Figure 5.1**    A line graph.

In line graphs, individual data points are connected by straight lines to reveal data trends. Line graphs use the Cartesian coordinate system, which was developed by French mathematician René Descartes (1596–1650). A two-dimensional Cartesian system consists of a horizontal axis (x-axis) and a vertical axis (y-axis). In a line graph, the x-axis is commonly used to denote time and the y-axis to denote quantity.

Figure 5.1 is an example of a line graph, showing the infant mortality rate in the US from 1935 to 2020. In this case, the data are captured every five years. Exact numbers (deaths per 1,000 live births) are indicated above the data points for easier reading. Figure 5.1 makes it clear that from 1935 to 2020 there has been a dramatic reduction in the infant mortality rate in the US. The reduction was most pronounced from 1935 to about 1985. From there on, the mortality rate continued to drop but the change was less dramatic, and after 2000 it has held stable.

The same data shown in Figure 5.1 can be displayed using a bar graph, as shown in Figure 5.2. Although these two graphs represent the same data, they create different visual impressions. The line graph highlights infant death rate change over time. That same change is present in the bar graph and indicated by

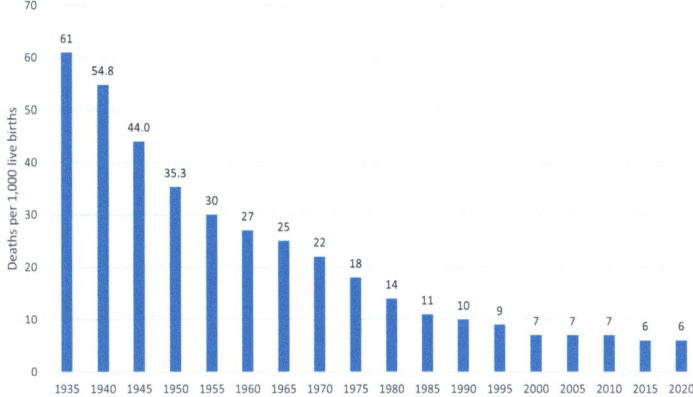

**Figure 5.2**    A bar graph.

the changing heights of the bars, but it is overshadowed by the individual bars. In other words, in bar graphs the primary visual emphasis is on the discrete data points – the death rate in each given year – rather than changes over time.

Sometimes, we see graphs that *look* like a bar graph, only without gaps between bars, such as shown in Figure 5.3. This graph shows the number of reported COVID-19 cases in the US in the early days of the pandemic (February–March 2020).

Figure 5.3 is, technically, not a bar graph but what is known as a histogram. A histogram represents frequency distribution, in this case the frequency of COVID cases. The difference between a regular bar graph and a histogram is that in a bar graph the *x*-axis plots discrete data. Discrete data are data that are countable and individually separate, like nationality, geographic region, or number of students in a class. Nationalities, for example, can be Japanese, French, or American, but there is no nationality *between* Japanese and French.

By contrast, in a histogram the *x*-axis plots continuous data. Continuous data are data that can take on an infinite number of possible values, such as height,

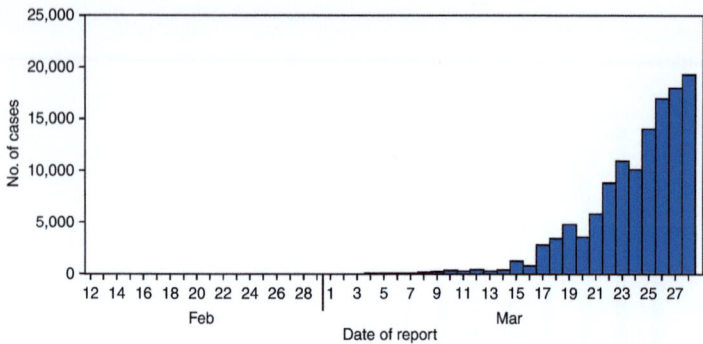

**Figure 5.3**   A histogram.

weight, or temperature. Temperature, for example, can be 20.17, 20.294, 21.6, etc. By removing the gaps between bars, histograms indicate this continuous nature.

Certainly, the difference between continuous and discrete data is not always clear-cut. Time, for example, is technically continuous, which is how Figure 5.3 treats it. However, if we break time into timeframes, it becomes discrete, which is how Figure 5.2 treats it.

Another difference between bar graphs and histograms is that in histograms the *size* rather than the *height* of the bar is used to indicate data. In other words, bars in a histogram may have different widths. Imagine graphing the results of a marketing survey in which participants' ages ranged from 18 to 28. Let's say we want to see all that data but are especially interested in the 20–25-year-olds. We can create a histogram and treat participants who are <20 as a single group, those who are 20, 21, 22, 23, 24, and 25 each as a separate group, and those who are >25 as a single group. The first and the last groups would have a wider bar than the others.

The last of the three common graphs is the pie graph, which breaks a whole set of data into multiple parts. "Pie graph" is a metaphor, comparing the graph to the common food pie, where individual slices add up to 100% of the pie. Figure 5.4 is an example, showing the different causes of deaths among

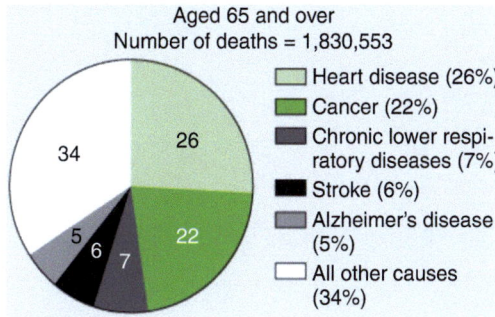

Aged 65 and over
Number of deaths = 1,830,553

Heart disease (26%)
Cancer (22%)
Chronic lower respi-
ratory diseases (7%)
Stroke (6%)
Alzheimer's disease
(5%)
All other causes
(34%)

**Figure 5.4**   A pie graph.

Americans 65 years and older in the year 2021. The sum of all deaths (the whole pie) is 1,830,553 deaths, which is broken down into six causes/slices. The slice representing heart disease is the largest slice representing a single cause, accounting for 26% of deaths, followed by cancer, chronic lower respiratory diseases, stroke, and Alzheimer's. All other causes are grouped together. As Figure 5.4 shows, pie graphs provide an at-a-glance, easy comparison of parts of the whole.

Besides line graphs, bar graphs, histograms, and pie graphs, occasionally we also see scatter plots in population science communication. Friendly and Denis attributed the origin of the scatter plot to English scientist Sir John Frederick William Herschel (1792–1871), who used a basic scatter plot to trace the orbits of stars in 1833.

Figure 5.5 is an example of a scatter plot, showing the correlation between percentages of adults with college degrees and the prevalence of diabetes. As the y-axis label indicates, the prevalence data are "smoothed," which simply means that the data were statistically processed to reduce outliers. In a scatter plot, each dot in the graph represents a separate "entity," which can be a person, a state, a country, etc. In Figure 5.5 each data point is a census tract (an area with an average population of 4,000) in King County, Washington (data were collected from January 2005 through December 2006).

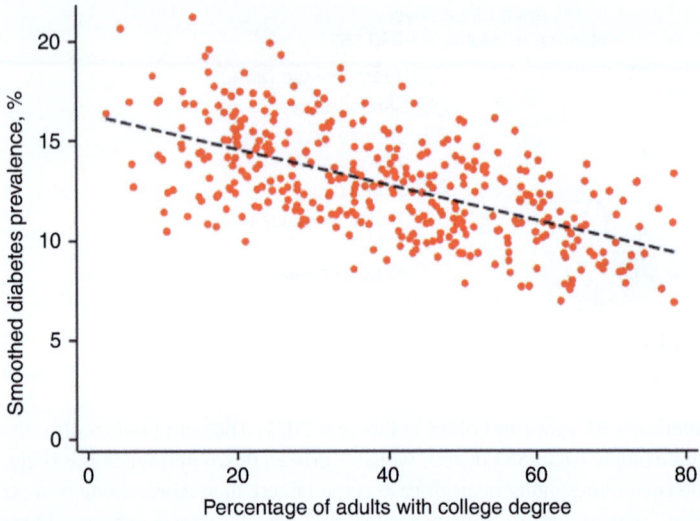

**Figure 5.5**    A scatter plot.

Each census tract is placed on the graph based on measurements from both the x-axis and the y-axis. Let's say that in one census tract 20% of the adults have a college degree. We can draw a vertical line at the 20-tick mark from the x-axis. Let's say in this census tract the prevalence of diabetes is 10%. We can draw a horizontal line at the 10-tick mark from the y-axis. The point where the two lines intersect is where we place this census tract. When all the census tracts (371 in Figure 5.5) are placed this way, the general pattern of the data can demonstrate the relationship between college education and the prevalence of diabetes in King County, Washington.

According to the data pattern shown in Figure 5.5, diabetes is less prevalent in census tracts where college degrees are more common. Conversely, diabetes is more prevalent in census tracts where college degrees are less common. In other words, there is a negative correlation between college education and diabetes. The dotted line, known as the line of best fit or the trendline, makes this relationship visually clear. The best-fit line is drawn *through* the data with about an equal number of data points on either side of the line having similar distances to the line. This way, it provides a visual approximation of the overall trend of the data.

## The Truncated Y-Axis

Now that we've met the most commonly used graphs in popular life science communication, we can move on to consider some of the misunderstandings and misconceptions surrounding them. A good place to start is the y-axis.

Conventionally, in a line graph, bar graph, or histogram, we put independent variables – such as time, geographic locations, names of presidential candidates – on the horizontal x-axis. We then plot dependent variables – such as sales revenues, prevalence of diseases, numbers of social media followers – on the vertical y-axis. This is how Figures 5.1–5.3 are structured. With this conventional structure, the y-axis presents the data that are being explained, the "point" of a graph, so to speak. Because of this, how we design the y-axis is tremendously important and can single-handedly change the story we tell with the data.

One notable pitfall with y-axis design is truncation, specifically in bar graphs, which use height to communicate quantities. Although this issue is well known in information design scholarship, it is worth repeating because the practice is still common and, as of this writing, is the default setting in Microsoft Excel when graphing data with small differences. A truncated y-axis, as the name suggests, is cut off, so it doesn't start at zero as the baseline, but rather at an arbitrary number. Although doing so doesn't, in and of itself, change the values of dependent variables, it can significantly change the overall visual appearance of the graph and thus readers' interpretation of the data.

Consider Figure 5.6, which presents two bar graphs of the same (fictional) sales revenue data. If shareholders were to look at Figure 5.6(a) they might be overjoyed by the significant revenue increase throughout the year, especially the dramatic increase in the last quarter. Unfortunately, reality isn't that fabulous. Figure 5.6(a), by starting the y-axis at $32,000, greatly exaggerates the sales growth. The actual growth is more accurately portrayed in

**Figure 5.6** The truncation effect.

Figure 5.6(b): still good, but not *that* good. There you have it, the so-called truncation effect – namely, when the *y*-axis is truncated, small differences are exaggerated.

Now, it is true that each bar in Figure 5.6(a) *is* labeled, and readers *can* examine them to ascertain the actual differences in data. But in the context of popular science communication (or mass media in general), how many readers *will* spend the time to scrutinize graphed numbers? And even if they do, global visual impressions trump specific numbers, so a truncated *y*-axis will still cause an overestimation of data difference. As Yang et al.'s studies showed, even when participants were explicitly taught the truncation effect immediately before viewing graphs, they still overestimated differences in truncated bar graphs.

Aside from truncation, other y-axis design choices can be problematic. For example, setting a high maximum bound on the axis can flatten differences between data and mask fluctuations. Choosing between percentages and actual values as the scale can also be tricky. As information designer Alberto Cairo recounted, a percentage-based y-axis once led to the rumored death of the web. In an August 2010 issue of *WIRED* magazine, an area graph (a form of line graph where the regions below the lines are filled with color) showed the shrinking percentage of internet traffic accessed through web browsers and the increasing percentages of traffic accessed through smartphones and other services. Given this data pattern, *WIRED* claimed that the conventional web was dying. Yet, if we use actual amount of internet traffic (e.g., in terabytes) as the y-axis, we will see that *both* web browser traffic and other kinds of internet traffic were increasing. Sure, more traffic was going through non-web-browsers, but the web was hardly dying.

## Logarithmic Scale (and What Is That?)

Truncation or percentage vs. total is hardly the most complicated issue when it comes to designing the y-axis. The logarithmic scale, I would venture, tops them all.

In a line graph (or bar graph, histogram, or scatter plot), the y-axis can assume either an arithmetic scale or a logarithmic scale. The arithmetic scale (also known as the linear scale) is by far the more common of the two. It is what's used in Figures 5.1–5.3 and 5.5. With this scale the same distance along the axis represents the same change in quantity. For example, in Figure 5.1 the distance between 0 and 10 (deaths per 1,000 births), 10 and 20, 20 and 30, etc. is the same.

Then there is the logarithmic scale. An easy way to think about "logarithm" is to think of it as the "exponent." For example, we know that $3^2 = 9$. In this equation, 2 is the exponent. This same equation can be rewritten as $\log_3 9 = 2$. In this equation, 2 is the logarithm. A logarithmic scale, then, shows exponential change. For example, the scale can start with 1 and then go to 2, 4, 8, etc.

Or, it can start with 1 and then go to 10, 100, 1,000, etc. With a logarithmic scale, the same distance on the axis represents the same *rate* of change.

The logarithmic scale is useful for presenting data that have a wide range. Consider the case of malaria. According to the World Health Organization's estimate, India has 15 million cases of malaria each year. In the US that number is about 1,700. The two numbers differ by several orders of magnitude. If we use the common arithmetic scale to plot the numbers, one or the other data point will be "off the chart," so to speak. A logarithmic scale solves this problem. By making the scale advance exponentially, it can reflect vastly different data.

During the COVID-19 pandemic, media outlets used both the arithmetic and logarithmic scales to track the spread of the disease and its damage. In Figure 5.7(a) we see COVID-19-related deaths in the US between February 15, 2020 and April 18, 2020 on an arithmetic scale; in Figure 5.7(b) the same data are plotted on a logarithmic scale. It makes sense that both scales were used. The arithmetic scale is familiar and easy to comprehend. The logarithmic scale, on the other hand, shows the exponential spread of the disease. Comparing these two graphs, we can see that the logarithmic scale better captures the increase in COVID-19 deaths in February and March when absolute numbers remained low.

The problem, however, is that nonspecialist readers tend to have difficulty understanding the logarithmic scale. It is uncommon and unfamiliar; it is not intuitive and involves numerical computation. Indeed, Romano et al.'s and Ryan and Evers' studies show that nonspecialist readers had less accurate understanding of how the COVID-19 pandemic was developing when the data were graphed on a logarithmic scale than on an arithmetic scale. This misunderstanding can have downstream impacts on health behaviors and beliefs.

As Figure 5.7 shows, when data are presented on an arithmetic scale we see a "skyrocketing" increase of COVID-19 deaths as time goes on; by contrast, when data are presented on a logarithmic scale, the line flattens. Both representations are correct: The *rate* of increase in COVID-19 deaths stabilized as the absolute numbers continued to climb. But if a reader doesn't fully understand what the logarithmic scale measures, the graph can create the visual

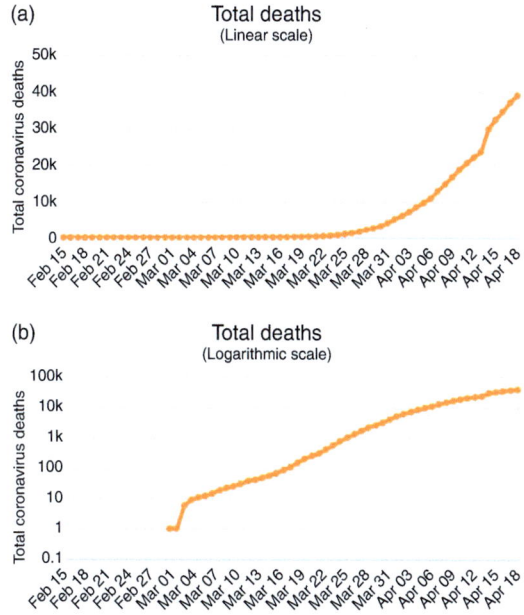

**Figure 5.7**  COVID-19-related death in early 2020 shown in arithmetic scale (a) and logarithmic scale (b).

impression that the pandemic is not *that* dangerous or devastating. Consequently, as Ryan and Evers found, people may become less supportive of policies and actions aimed at reducing COVID-19, such as social distancing and mask-wearing.

Changing the scale changes the way numbers look. The logarithmic scale may be better at capturing exponential growth, but if people have trouble understanding it, it should not be a preferred choice in popular communication. At least, it should not be used without careful explanation. Presenting the same data side by side using both the familiar arithmetic scale and the logarithmic scale may also help to communicate the full extent of the data.

## Interpreting the Scatter Plot

Scatter plots are commonly used in scientific literature. In fact, some scholars think that scatter plots are the most versatile and useful of the statistical graphs. In popular science communication, scatter plots are less common compared to line, bar, and pie graphs, making their conventions less of a common knowledge. And, by their very nature, scatter plots *are* more complex because they are not simple comparisons of data across time or categories. Instead, they use pairs of numerical data to look for relationships. For us to continue using scatter plots in popular science communication, we need to ask "do people actually understand them?"

In a 2014 Pew Research Center survey, one of the questions concerns the scatter plot. The question presents the scatter plot shown in Figure 5.8 and asks survey participants which of the four statements best describes the data shown: (A) In recent years, the rate of cavities has increased in many countries; (B) In some countries, people brush their teeth more frequently than in other countries; (C) The more sugar people eat, the more likely they are to get cavities; and (D) In recent years, the consumption of sugar has increased in many countries.

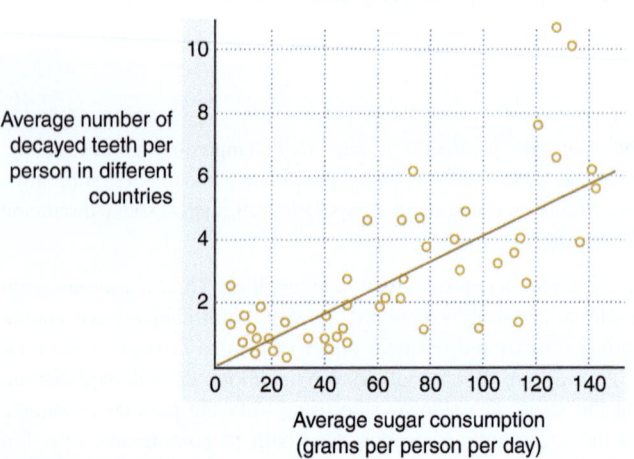

**Figure 5.8**    Scatter plot used in the Pew Research Center survey.

According to Pew, the correct answer is C and 63% of participants got it right. Education had a significant impact on the result. While 79% of people with a college degree and 84% of people with a postgraduate degree chose the correct answer, among people with a high school degree or less, only half gave the correct answer.

Are these results good enough? I guess that depends on who you ask. On one hand, more than half of the people apparently chose the right answer. On the other hand, a significant percentage did not get the gist of the information. If we want to engage all sectors of the public in science, especially people who are not already "fans of science," then the 50% accuracy rate among people without a college degree is concerning.

In addition, something else about the Pew survey bothers me: The supposedly correct answer C sounds dangerously close to a causal statement: the more x happens, the more y follows. No, the statement doesn't actually use the word "cause" or "because," but in everyday language use causal claims are often made implicitly. The implicit causal premise here is what linguists call co-occurrence: x causes y because x frequently occurs together with y.

Yet, scatter plots do *not* suggest causation, only correlation (or connection or association). The correlation can be positive, which means that x and y increase together. Or, the correlation can be negative, which means that when one increases, the other decreases. A scatter plot makes no claim about what caused what. In the case of sugar consumption vis-à-vis cavities, it could indeed be that higher sugar consumption led to increased cavities. Or (indulge me here), it could be that people with more cavities, warned by dentists to reduce sweets consumption, psychologically find sugar irresistible and wind up consuming more of it. Or it could be that some other factors caused both. A better answer to Pew's survey question, then, should be "There is a positive correlation between sugar consumption and numbers of cavities."

I wonder if we substitute this statement (which sounds vague) for the original C statement (which sounds concrete and makes common sense), how many participants in the survey would have chosen it. I also wonder how many participants who chose C assumed that sugar consumption *caused* cavities.

Then, there are important, finer details about scatter plots that the Pew survey did not test. In a scatter plot, the correlation between *x* and *y* can be strong or weak. If data points cluster tightly around the best-fit line, then the correlation is strong. If data points are scattered and distant from the best-fit line, the correlation is weak. In the example used by Pew, the correlation is not *that* strong, and there are clear outliers, especially in the top right corner. I wonder if and how many participants noticed that and what they thought of that.

When data points are *really* scattered every which way, then there is no real correlation between *x* and *y*. However, if you ask for it, graphing software like Excel will still give you a "best-fit" line that runs haphazardly through the data, as Figure 5.9 shows. If we want to be tricky, this graph would make a good survey question to test if readers can properly interpret scatter plots.

Unfortunately, I could find no formal studies that tested these fine-tuned understandings of scatter plots. Curious, I ran a rather informal poll on social media, asking friends in my social circle what they think the diabetes–education scatter plot in Figure 5.5 means.

I received 72 responses (to put this into perspective, at the time of the poll I had over 600 contacts in my network). These responses are, of course, not at all

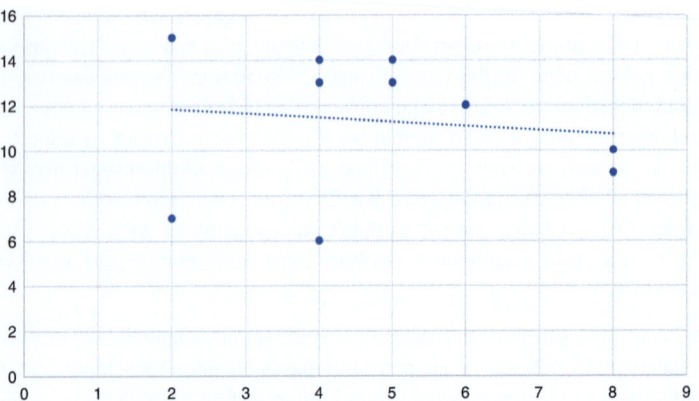

**Figure 5.9** A bogus best-fit line.

a representative sample. People who elected to respond are, more likely than not, people who already had (or thought they had) a decent understanding of the graph. For what it is worth, though, I find the results interesting.

A little more than a dozen indicated that they don't quite know what to make of the graph. The rest commented, correctly, that it depicts a relationship between education and diabetes. Only about a handful of these made explicit causal statements, such as "if you have a college degree, you have a lower chance of diabetes" or "more education results in less diabetes." In several other cases, the respondents protested the data in a way (stating that the data are not "reliable") that shows they *thought* the graph claimed causal relationship. Nineteen responses specifically used the words "correlation" or "association" or otherwise denied a causal relationship between education and diabetes. The rest gave the same kind of response that the Pew survey used, where it is difficult to tell, based on the statement alone, whether the person was thinking of correlation or causation. Finally, a handful of people commented on the "dotted line," but none was quite right in explaining what it means. The common response was that it depicts the average or the median (the middle number in a sorted list of numbers) – when it is a trendline of the overall data pattern.

Because scatter plots show individual "raw" data, their use in the mass media is a welcome sign that we are engaging the public in the process of interpreting research findings. It is a gesture of transparency. That said, we desperately need formal studies on how the public understand scatter plots – or *if* they do. Findings from such studies will help us to use (or not use), design, and explain scatter plots in ways that maximize accessibility. As of now, there is a dearth of formal studies on one hand and a nervous attempt to use scatter plots on the other, which can result in confusing graphs such as Figure 5.10. This graph was designed based on a scatter plot published by ABC News (data in the figure are illustrative, not precise, taken from Mitropoulos et al.'s study of COVID vaccinations).

Figure 5.10 shows the correlation between COVID-19 vaccination and case load in different states in the US. As mentioned earlier, in a scatter plot, each data point is the intersection of two measurements. So, in this case, each dot should represent a state, and its placement on the graph should reflect both its

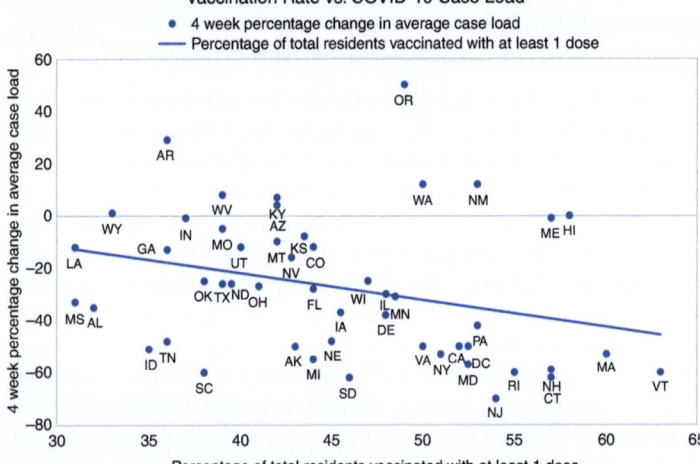

**Figure 5.10**   A confusing scatter plot.

COVID-19 vaccination status and case load. Yet, the graph explicitly, and mistakenly, labels each dot as the case load and the best-fit line as the vaccination status. These design features make an already unfamiliar graph even more confusing and contribute to misconceptions about the scatter plot.

## Variations of the Bar

The simple bar graph shown in Figure 5.2 has several variations, not all of which are equally easy to read. The most common variation is the horizontal bar, which is the same bar graph rotated 90 degrees so the bars extend horizontally (Figure 5.11). This design is useful when the labels for individual bars are long and difficult to fit under the horizontal scale. Listing them vertically on separate lines creates a clearer presentation. Visually speaking, the horizontal bar graph is just as easy to read as the vertical bar graph. Instead of comparing the bars across the top on the y-axis, we compare them on the right edge on the x-axis.

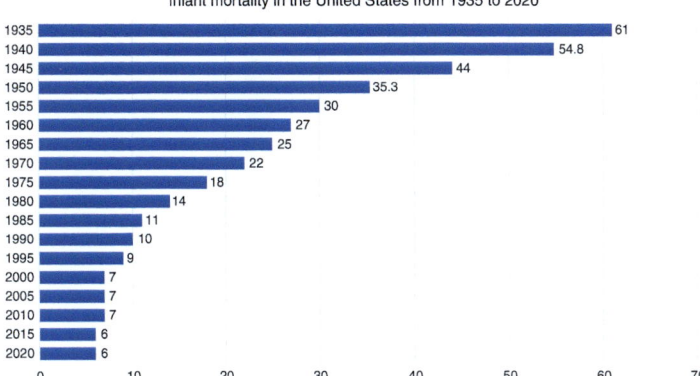

**Figure 5.11**   A horizontal bar graph.

Both Figures 5.2 and 5.11 compare data across one category ("number of deaths"). When we try to compare data across multiple categories, we have the so-called grouped bar graph. Figure 5.12 is an example. It shows the percentages of Americans with HIV who are at different stages of care: diagnosed, linked to care, retained in care, prescribed antiretroviral therapy (ART), and virally suppressed (meaning the amount of virus in the body is reduced to an undetectable level).

Figure 5.12 highlights the advantage of bar graphs at making multiple points of comparison at once. Not only does it compare data across the five different stages of care, it also compares data across racial/ethnic groups. That said, it is somewhat difficult to visually compare percentages of people at the same stages of care across racial/ethnic groups because readers have to bypass additional bars and the gaps between them.

More complicated than the grouped bars are the stacked bars. Like grouped bar graphs, stacked bar graphs show multiple categories of comparison, but they stack the subcategory data into a single bar. This is shown in Figure 5.13, which compares poisoning deaths involving opioid analgesics, other drugs, and no drugs in the US from 1999 to 2006.

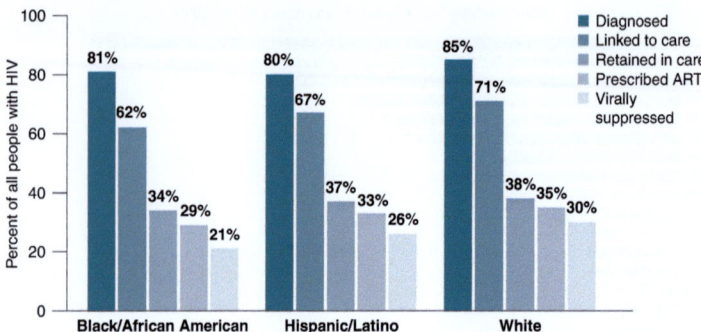

**Figure 5.12**   A grouped bar graph.

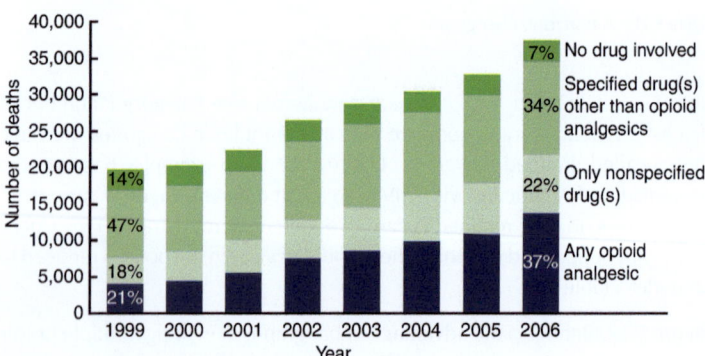

**Figure 5.13**   A stacked bar graph.

Stacked bars are a bit of a controversy because they have distinct advantages but also distinct problems. They allow comparisons of total sums of data. That is, in Figure 5.13 it is easy to see that from 1999 to 2006 total poisoning deaths steadily increased. It is also easy to see the change over time in the "any opioid analgesic" subcategory. This is because this set of bar segments all start at the x-axis and have a common baseline. Therefore, we can compare them by judging their relative positions off of the y-axis, which is an easy perceptual task.

However, to visually compare any other subcategories, such as "no drug involved," across the years would be difficult. As Cleveland and McGill's and Talbot et al.'s experiments show, these bar segments don't have a common baseline, so we have to gauge their individual *lengths* for comparison, which is a difficult perceptual task. Similarly, to visually compare the different causes of poisoning *within* a year is difficult.

For these reasons, people like Kosara argue that stacked bars are "terrible": They have the appearance of data richness but don't actually convey a whole lot of data. Others, by contrast, argue that stacked bars have their place because they can compare category totals and the changes that are happening in at least one subcategory (and possibly more if the change is obvious).

So, whether or not to use stacked bars is a bit of a judgment call and depends on what we want to emphasize in a given set of data. At the very least, we need to be very aware of their advantages and disadvantages and assess their fitness for purpose.

Finally, there is the cousin of the stacked bars: the 100% stacked bars. In a 100% stacked bar graph, all of the bars are pulled to the same height to equal 100%, and the subcategory data are converted to percentages. Figure 5.14 is an example. It shows the distribution of COVID-19 vaccines across US counties from December 14, 2020 to March 1, 2021.

Because COVID-19 disproportionally affects people who are economically and socially disadvantaged, the CDC wants to track whether vaccine distribution equitably reaches the most vulnerable populations. To gather these data, CDC ranks all counties based on the social vulnerability index (SVI), which includes 15 indicators. These indicators are in turn categorized into four themes: socioeconomic status (indicators 1–4), household composition and disability status (indicators 5–8), racial/ethnic minority status and language (indicators 9 and 10), and housing type and transportation (indicators 11–15). For each indicator, theme, and the overall SVI, a county can be ranked as having high vulnerability, moderate vulnerability, or low vulnerability to COVID-19. The graph then proceeds to show vaccine distribution across the three vulnerability groups.

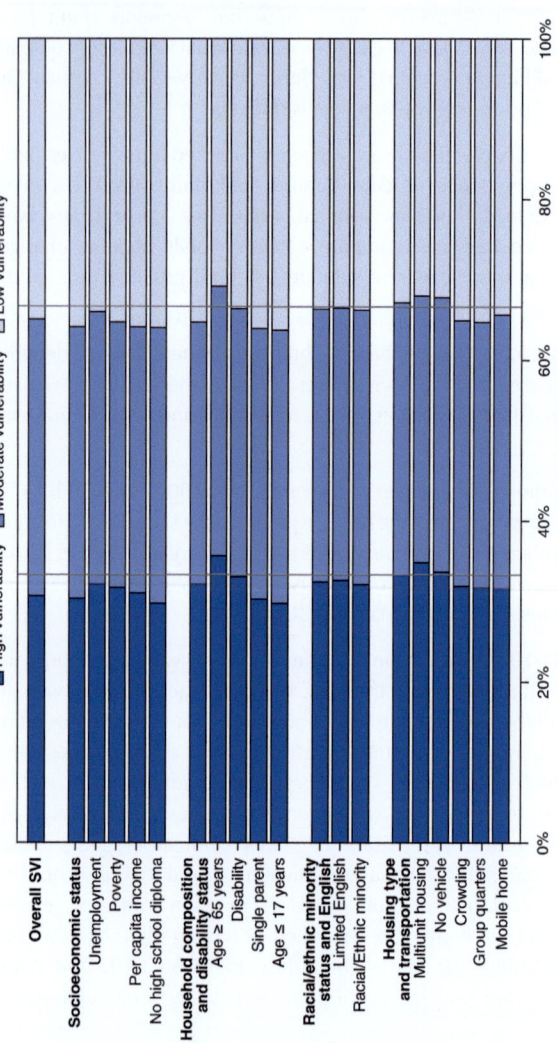

**Figure 5.14** A 100% stacked bar graph.

Compared with a stacked bar graph, a 100% stacked bar graph provides two baselines: one at the top and one at the bottom (if the bars are vertically placed), or one on the left and one on the right (if the bars are horizontally placed, as in Figure 5.14). Technically, then, two subcategories of data can be easily compared. In Figure 5.14 that would be the high-vulnerability group and the low-vulnerability group. That said, segments that are in the middle would still be difficult to compare.

Figure 5.14 attempts to solve this problem by adding vertical baselines *inside* the graph, separating the bars into three equal lengths of one-third. Still, at a glance, the graph looks visually busy and the conclusion that vaccine distribution is higher in low-vulnerability counties and lower in high-vulnerability counties is not immediately clear.

As with stacked bars, we need to carefully weigh the pros and cons of the 100% stacked bars and their fitness for individual contexts before using them.

## Excessive Details

We have all heard of the saying "less is more." In graphs, that saying rings true in more ways than one. The challenges with stacked bars that we see above stem precisely from our attempt to cram too much data into a single display. The details end up becoming confusing.

Certainly, not all details are created equal. Some of them are little more than visual decorations, for example, the use of three dimensions, as in bar graphs or pie graphs where each bar or pie slice is a three-dimensional object. These graphs can easily hamper communication: the bulky bars may obscure each other, the top of the bars with tilted planes can be difficult to read off of the axis, and the sizes of three-dimensional slices are difficult to visually compare. The dimensions are added to make graphs appear as if more professional and authoritative, as opposed to being more informative and accessible.

Then, there are superimposed clip art or other pictorial images. Figure 5.15 is an example, a line graph showing Americans' declining consumption of milk. If the subject matter of the graph is something readers are unfamiliar or less familiar with (let's say Americans' consumption of the spotted seatrout), then pictorial elements may be justifiable as they can convey the visual context of

## Milk's Massive American Decline
Per capita consumption of milk in the U.S. (in pounds per person)

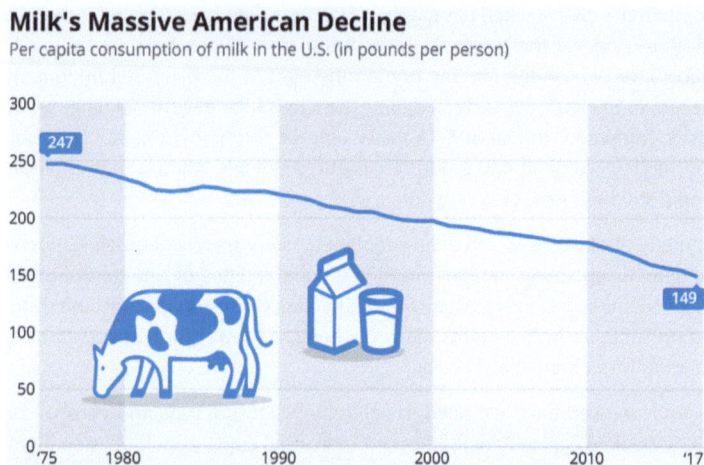

**Figure 5.15**   A line graph with unnecessary pictorial decoration.

what the spotted seatrout looks like. But in the case of milk, they seem quite unnecessary. They are added to make the graph appear more interesting, as opposed to being more informative.

The good news is that, for the most part, designers of graphs are aware of the drawbacks of decorative details. We owe much of this awareness to renowned data visualization theorist Edward Tufte, who calls purely decorative details (which also include things such as dazzling cross-hatching patterns or excessive colors) non-data-ink or redundant data-ink. For Tufte, graphs that use non-data-ink are chartjunk. Understandably, no one wants to be accused of creating junk!

Less heeded, it seems, is the danger of another kind of excess: data-ink, lots and lots of data-ink. This kind of excessive detail, I argue, can cause more problems in popular science communication. Consider, for example, Figure 5.16, which illustrates the numbers and percentages of deaths caused by pneumonia, influenza, COVID-19, and a combination of these conditions (known as PIC) in the US between October 2016 and October 2020. The

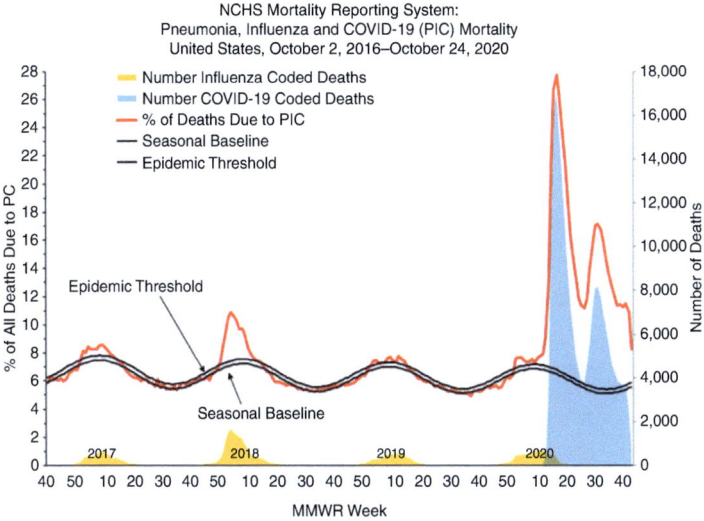

**Figure 5.16** An overly complicated graph.

graph combines a regular line graph and an area graph. With the many data it attempts to measure, Figure 5.16 sports three data lines, five data regions, and two different *y*-axes. Readers must figure out which *y*-axis they need to use for which data, a process that involves glancing nervously back and forth between the data line/region, the key, and the two *y*-axes. To make things worse, the *x*-axis measures "week," but some of the data regions are labeled in years, challenging readers to reconcile the difference. It also doesn't help that the graph gives no clue what MMWR is, what epidemic threshold is, or what seasonal baseline is.

In short, Figure 5.16 is overly complicated. All of the ink conveys information, and the graph is probably quite enlightening to a specialist audience who can mentally process all of the data and appreciate having them co-located in one graph. But for nonspecialist readers who do not have an intimate knowledge of the subject matter and are less familiar with graphing conventions, Figure 5.16 looks confusing and reinforces the belief that science is esoteric and

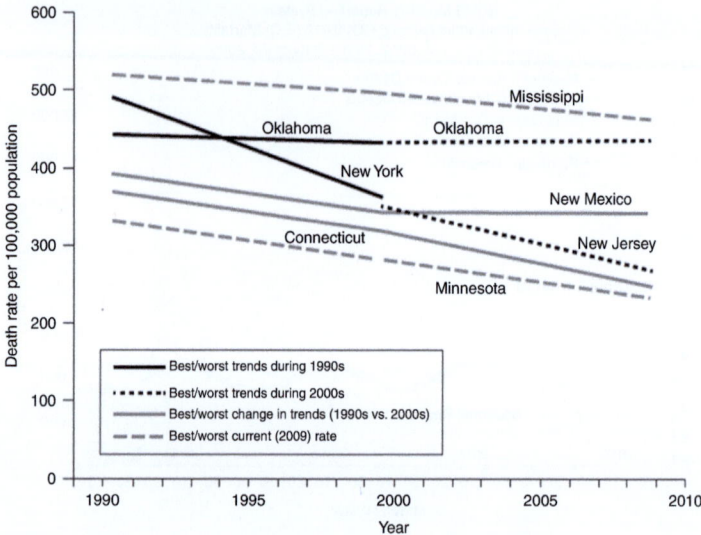

**Figure 5.17**  Another overly complicated graph.

inaccessible to everyday readers. And if science is inaccessible, the idea goes, there is no stopping scientists from manipulating data to tell whatever stories they desire. In other words, overly complicated graphs may inspire neither public understanding of science nor trust in science.

Another example of an overly complicated graph is shown in Figure 5.17, this one illustrating the rates of premature (less than 75 years old) death across US states in the 1990s and the 2000s. The graph attempts to show which two states experienced most reduction in premature death rate in these two time periods (i.e., the best trends) and which two states experienced the most increase in premature death rate (i.e., the worst trends). It also attempts to show *changes* in trends between the two time periods: a negative change means an improvement in death rate trends; a positive change means a worsening in death rate trends. So, the largest negative change would be the best change in trends, while the largest positive change would be the worst change in trends. Finally, the figure shows the states with the best (lowest) or worst (highest) current (as of 2009) death rate.

With all these ambitions, the graph ends up with eight data lines and data labels, which is quite a few for a line graph. It is also very confusing that the same style of lines, such as solid black lines, can stand for *both* the best and the worst in the same category, and readers are expected to decide which is which based on data trends. Add to this the inherently complex and abstract nature of the data measured (rates, trends, changes in trends), and this line graph is quite inaccessible to everyday readers.

Tufte states that excellent graphs should give readers the greatest number of ideas in the shortest time and smallest space with the least ink. He is, of course, right. But information richness is not the only thing that matters in popular science communication. Everything being equal, the richer the data, the higher the chance of overwhelming readers' short-term memory – and humans' short-term memory is notoriously limited. For nonspecialist readers it may be more advantageous to be a little less economical. That is, rather than jamming so much data into a single, overly complicated display to save time, space, and ink, we might want to use multiple but simpler graphs to present the data – if we decide that all the data are indeed essential to graph.

## Pictographs

Tufte's charge of chartjunk also puts a general fear into graph designers about pictographs, which use pictorial images to illustrate quantitative data. In his work, Tufte gave multiple examples of poorly designed graphs with pictorial images: using dollar bills of different sizes to chart the rate of inflation, or using oil barrels of different sizes to graph the changing price of oil. The problem with these graphs is that they use two-dimensional (or even three-dimensional) images to represent numbers, which are only one-dimensional. Doing so is numerically tricky, as a designer needs to do the math of converting one-dimensional data into multi-dimensional images (for more about this math, see Chapter 7). Even when this is done right, it is difficult for readers to visually compare the data: Our eyes and brains are good at judging positions (such as the location of a bar on a scale) but less so at judging area or volume.

China

India

United
States

Brazil

Pakistan

Each ![icon] = 1,000,000 metric tons

**Figure 5.18**   A pictograph.

However, actual pictographs work fundamentally differently from these examples. Also known as icon arrays or visual tables, pictographs use icons such as human figures, everyday objects, and simple geometric shapes to indicate quantitative data. In pictographs, the *number*, not the *area* or *volume*, of pictorial images encodes data. That is, a single icon is a base unit, which can denote a number of 1, 10, 100, or whatever it needs to be. Two icons, then, represent the number 2, 20, 200, and so on. In viewing pictographs, the visual task being performed is not one of comparing distinct area or volume but one of comparing the amount and saturation of data-ink.

Figure 5.18 is an example, showing the top five cotton-producing countries worldwide in 2020/2021 and their respective cotton output. Each cotton icon represents 1,000,000 metric tons. Multiple or partial icons scale the number up and down accordingly. By lining up the icons for different countries, the graph gives an immediate and concrete visual impression of how cotton production varies among the top five.

Certainly, Figure 5.18 as designed doesn't give readers the precise data. To remedy that, we can write out the numbers at the end of each row (6,423,000 metric tons, 6,162,000 metric tons, etc.). It is useful to note that the same issue is faced by other graphs such as bar graphs. Without labeling the bars, we will be hard pressed to read exact numbers off of an axis. It is also useful to note that, in popular science communication, it is often not essential to provide precise numbers. While a scientist audience may need to know that the data in question is 236.5 in order to replicate a previous study or verify experimental results, nonspecialist readers only need salient and significant data trends and patterns to make sense of scientific findings. In the words of the International System of Typographic Picture Education (Isotype) initiative (as quoted by Burke), "To remember simplified pictures is better than to forget accurate figures" (p. 215).

The Isotype initiative was spearheaded by Otto Neurath (1882–1945), an Austrian philosopher and sociologist, in the 1920s at the Vienna Museum of Society and Economy. The initiative was developed to present social economic data for everyday citizens. An example of Isotype is shown in Figure 5.19, which illustrates the number of people immigrated into and emigrated from various European and North/South American countries between 1920 and 1927.

In Figure 5.19, each human icon carrying a suitcase represents 250,000 people. Icons that are facing right and walking away from a country are emigrating out of that country. Icons that are facing left and walking toward a country are immigrating into that country. The emigrants start walking away from a vertical line in the middle of the graph; the immigrants pick up where the emigrants leave off. This way, the vertical line serves as a population break-even point. If there are icons to the left of the line, then there is a surplus in immigration. Otherwise, a country is losing population.

Standalone pictographs such as Figures 5.18 and 5.19 are not all that common in popular life science communication. Part of this, as mentioned earlier, may be designers' fear of being called out for creating chartjunk. Another reason may be the technical difficulty of producing pictorial images. Yet another reason could be the stereotype that pictographs are suitable only for children and nonserious communication.

But I believe pictographs are a missed opportunity for popular science communication because they offer multiple benefits. Most obviously, compared with conventional Cartesian graphs, the use of pictorial images adds visual

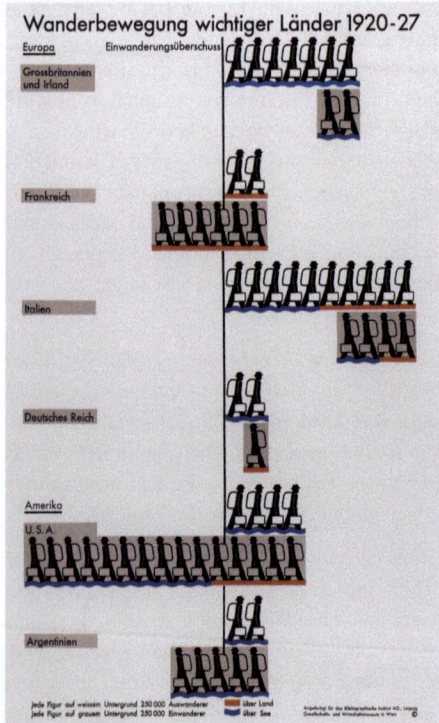

**Figure 5.19**    An Isotype example: movement of people across countries, 1920–1927.

interest and attracts viewer attention. As we have mentioned multiple times in earlier chapters, attracting people's attention is a prerequisite for engaging them in science communication.

But this is not all. Pictographs also offer distinct advantages when it comes to communicating data. First, when we use pictorial images that are indicative of the nature of data, such as human figures for human populations, data become more intuitive, more concrete, and thus easier to grasp. Second, pictographs are well positioned to tell sophisticated stories about data. Figure 5.19, for example, tells whether a country is gaining or losing population, which country has gained the most or lost the most population, and how different countries compare in their immigration/emigration patterns. Moreover, each human figure stands on a small line, which is either a brown straight line or a blue wavy line. This line is coded with extra information: Those figures on a brown straight line traveled by land, while those on a blue wavy line traveled by sea.

One place where we see the potential of pictographs being picked up is medical and risk communication. Quantitative data are frequently used in these communication contexts to help patients understand treatment effects and risks. These data can be difficult for patients and even medical staff to interpret and discuss. Pictographs, as Fagerlin et al. and Garcia-Retamero et al. – as well as other studies – show, present a promising solution. They can help to highlight numerator information, which is the "foreground" data (such as how many people responded to a treatment), as well as denominator information, which is the "background" data (such as how many people in total received treatment). Pictographs can help reduce the effect of anecdotal experience on medical decision-making. Some people also find pictographs using human icons easier to identify with than standard graphs.

These benefits can be seen in Figure 5.20, which shows the result of a hypothetical drug trial. Out of 100 women (or 1,000, or 3,000, depending on how we define the base unit) who took an experimental drug, 35 (blue) had a beneficial effect, 9 (red) had adverse reactions, and the rest (gray) had no response. Figure 5.20 provides an at-a-glance, intuitive pattern of the data that helps people to holistically answer the question "how effective is this drug?"

Certainly, not all life science research data are about concrete entities such as humans or cotton. When pictorial images don't apply, simple geometric shapes such as small squares can be used to encode data. Garcia-Retamero's studies show that with the use of such shapes, pictographs retain their positive effects.

## Conclusion

Graphs are an essential tool for communicating numerical data, which are plentiful in life science research. Commonly used graphs such as line graphs, bar graphs, and pie graphs should be easy to understand, provided that the

**Figure 5.20**    Pictograph of hypothetical drug trial results.

dependent variable scale is ethically and carefully designed and that the graphs are not burdened with superfluous or relevant but excessive details.

At the same time, bar graphs do have multiple variations. Some of these variations are inherently complex for visual processing and should not be used without careful deliberation of their pros and cons for a given case.

Less familiar graphing conventions such as the scatter plot or the logarithmic scale risk confusing nonspecialist readers or convincing them that science is an inaccessible enterprise. The same goes for cramming excessive data into a single graph. The effect is one of alienation, not engagement.

Graphs that focus on salient trends and patterns, that rise above a rigid concern with data precision or economy, and that intelligently use pictorial images offer some of the ways to make numerical information in the life sciences accessible to the public.

# 6   Interactive Visuals

Contemporary life sciences are big data sciences. The human genome, for example, contains about three billion DNA base pairs and an estimated 20,000 protein-coding genes. Public health data, as another example, are endlessly enormous and encompass electronic medical records, health monitoring data, environmental data, and more. When it comes to analyzing and presenting these big data, interactive online visuals – maps, graphs, three-dimensional models, even computer games – have inherent advantages. They are dynamic and easily updated. They support user interaction and allow users to create displays that make sense to *them*. Being "hands-on" also makes these visual displays more interesting. As computer visualization technologies continue to advance, we are guaranteed to see faster, more fluid, more ingenious interactive displays.

That said, just because something is online and interactive or uses the latest technologies does not necessarily mean it is going to be engaging and meaningful. As we will see in this chapter, when it comes to computerized visuals there seems to be a prevailing assumption that what works for experts will automatically work for the public. The idea goes that experts have "higher" demands (visuals with more data, more functions, more interactions), while the public have "lower" demands, so if we aim high, the lower end will take care of itself.

This assumption, as I argue in this chapter, is false. Advanced technology does not equate to effective communication. Not only do interactive visuals need to be usable – for example, a button is easy to press and when pressed will result in a certain action – they also need to be useful. That is, they need to support the kind of actions and results that users find meaningful. And what is meaningful to a specialist audience is not going to be the same as what is meaningful to a nonspecialist audience.

## Interactive Maps

During the COVID-19 pandemic, interactive maps were commonly used to track data related to cases, deaths, hospitalization, and vaccination across countries and the globe. The US Centers for Disease Control and Prevention (CDC), for example, housed US maps with both state-level and county-level statistics. The Johns Hopkins Coronavirus Resource Center likewise housed both US and global maps. These interactive visuals allowed the display of up-to-date, geographically coded data in a familiar map format.

One of the most basic features and advantages of these interactive maps is the ability to instantly access location-specific data. Generally, this feature works with a simple mouse-over. That is, when users put their computer mouse cursors somewhere on the map, a small pop-up window will appear that provides the name of the location and the data in question. This function can be seen in Figure 6.1, which shows the number of COVID-19-related emergency department visits and percentage changes by US states and territories (territories appear as boxes below the map, not actual locations on the map).

A second common feature and advantage of interactive maps is the ability to filter and present different kinds of data. With Figure 6.1, users can choose from the options at the top of the map to view different data (hospitalizations, deaths, etc.) in the same map format.

Johns Hopkins' "The Demographics of COVID" map illustrates this advantage more fully. To understand how COVID-19 impacts people from different social groups, Johns Hopkins provided a series of maps where users can see the breakdown of pandemic impact among people of different age, gender, sex, race, and ethnicity. For example, by first selecting "Race" and then "Black or African American," viewers can see four maps showing, respectively, the accumulation of COVID-19 cases, deaths, tests, and vaccines among the Black/African American communities in different states across the country. Similarly, by first selecting "Age" and then "60–69," viewers can change the four maps to show corresponding data.

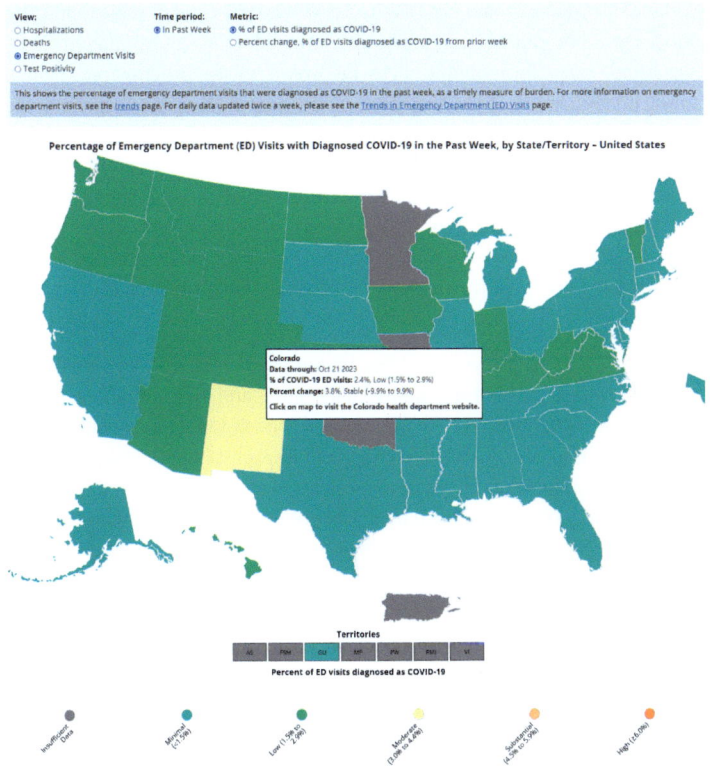

**Figure 6.1**    Interactive COVID-19 map, by US state and territory.

A third common feature of interactive maps is the use of color-coding to denote data. Typically, the lighter the color, the lower the number. For example, 0–15,000 cases per 100,000 population is colored the lightest shade of pink, 15,001–20,000 cases per 100,000 population appears slightly red, and so on. This feature allows an at-a-glance view of how different locations compare, where the "hot spots" are, and where the "safer areas" are.

This kind of color-coding is not new and is commonly used in static maps. But online, the color can become dynamic. Johns Hopkins' "Cumulative Cases Over Time" map, for example, keys the colors to the progression of time. Starting on January 23, 2020, the entire globe (except for China) starts with the lightest shade of yellow, representing zero cases. As time goes on, China quickly darkens. Then other parts of the world follow: the US, France, Canada, Italy, Australia, South America, Russia, all over Europe, all over Asia, all over Africa. If readers want to jump to a certain time, they can use the slider at the bottom of the map to move forward or backward. This kind of animation and interaction would not be possible with static maps.

Despite their advantages, interactive maps are not without potential issues. In Figure 6.1, locating individual states or territories is a relatively easy task because the sizes of these regions are fairly large and pinpointing them with the mouse is straightforward. In other cases, this task becomes challenging. For example, the CDC's "COVID-19 Vaccinations by County" map provides various vaccination data (primary series, booster doses, etc.) across all US counties. Since some states have over 100 counties, the map contains a large number of tiny regions. Directly clicking on any county is much more difficult.

To resolve this, the CDC provides two drop-down menus that allow users to first choose a state and then a county from a corresponding list. But this function requires scrolling through a long list of options. An arguably more user-friendly feature is the ability to search for specific locations, which Johns Hopkins' "COVID-19 United States Cases by County" map allows. Keying in the name of a town or city in the search box on top of the map will zoom a user to the location. Clicking on the area will bring up associated county-level data.

The challenge of viewing location-specific data is not just about designing more efficient technologies or more user-friendly features. At the bottom of it are two competing attempts: the attempt to provide users with large, complete sets of data, and the attempt to make that data meaningful to individual users. Epidemiologists may be interested in detailed, community-level data across the US, but most residents are likely to be interested in what's happening in their local communities. If having more data compromises the ability to individualize data, then more is not necessarily better.

It is not just the number of locations that can cause complications. The Johns Hopkins county-level case map, even though it facilitates location search, becomes overwhelming because of the sheer number of choices and amount of information it provides. Clicking on a county brings up a pop-up window (Figure 6.2) that shows the number of COVID-19 cases per 100,000 population in that county. The pop-up window also contains an infographic with additional data: the breakdown of cases by age, race, ethnicity; facts about healthcare and insurance; etc. Because the pop-up window is tiny, readers have to click and enlarge the infographic to be able to read it. There are additional options on the pop-up window that one can click. "Zoom to" has an obvious meaning and, when pressed, does what one expects (zooms into the map). "Pan," when pressed, moves the map very slightly with no obvious, intentional effect. "Select" sounds ambiguous and again gives no apparent visual results when pressed.

Above the pop-up window is a drop-down menu that, when clicked, displays several options (Figure 6.3(a)). Despite spending quite some time playing with the options, I'm not clear what they accomplish. It seems that by selecting one option, say "circle," one can then use the computer mouse to draw a circle on the map, which brings up yet another button with numbers on it (Figure 6.3(b)), but the function of that is likewise elusive. The map interface itself also has more buttons (Figure 6.3(c)). In addition to the useful "search" button mentioned above, there is a "home" button that brings one to the

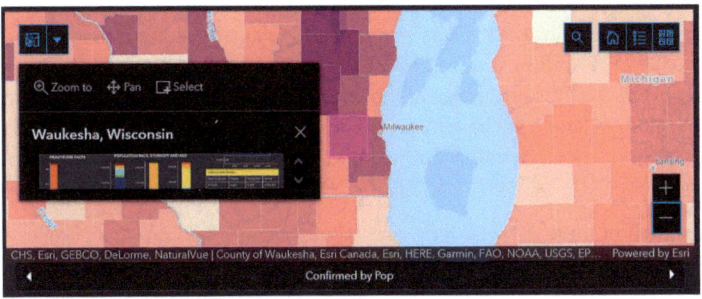

**Figure 6.2**  Johns Hopkins COVID-19 case map, by US county.

(a)

(b)

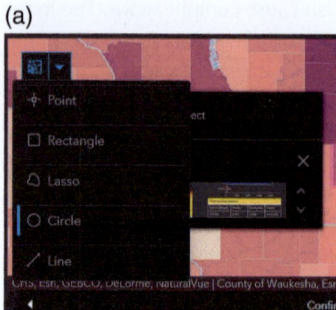
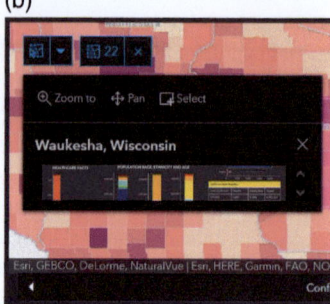

(c)

**Figure 6.3**   Additional options in the Johns Hopkins COVID-19 county case map.

"native" state of the map; a "list" button that explains the color legends of the map (i.e., light yellow is 0–17,255 cases); and a "squares" button that allows one to change the appearance of the base map. Apparently, one can choose from almost 30 different base maps, from a Streets view, to an Oceans view, to a Firefly Imagery Hybrid view, to a Mid-Century Map view and many, many more.

Given all of these data, buttons, and menus, there seems an implicit desire among visual developers to layer ever more information, more interactions, and more choices into the maps. More, it seems, is assumed to be better – because the maps can locate and filter data. But, at some point, too many choices will complicate the user interface and paralyze users. Just because we can do it doesn't mean that we should. Scrolling through 30 choices is far more difficult than scrolling through three. Similarly, less-than-intuitive inter-actions take time to figure out – if one ever figures it out. Finally, is that added labor necessary? What, for example, is a "Firefly Imagery Hybrid" map? Why do we need to layer COVID-19 data onto a mid-century map?

Aside from confusing and overwhelming users, another serious issue with overly complicated maps is that they demand quite a bit of bandwidth. The Johns Hopkins county-level case map is fairly slow to load even with a stellar internet connection. When internet connection is less than stellar, user frustra-tion will mount.

I want to flatter myself to think that my critique above is fairly on-target, because Johns Hopkins' developers have removed many of these options in the latest version of the map.

## Interactive Graphs

Interactive graphs, even more so than interactive maps, surged online during the COVID-19 pandemic. Enormous epidemiology data were daily collected around the world, and health agencies and news media like CDC, Mayo Clinic, *The Guardian*, and *The New York Times* curated these data and displayed them in various graphic formats.

As with interactive maps, a common feature and advantage of interactive graphs is the ability to easily read data. Readers do not need to use eyes or fingers to trace data off of the axis. Instead, a line graph can be configured so that when readers put their mouse cursors anywhere on the data line, a small pop-up window or side bar appears and displays the precise reading of the data at that point. Similarly, in a bar graph or histogram, when readers put their mouse cursors on a bar, the data in question appear. Figure 6.4 is one example.

Another common feature and advantage of interactive graphs is the ability to toggle between different kinds of data or combine multiple sets of data. Take Figure 6.4 as an example. In its present view, the bars show the numbers of cumulative COVID-19 deaths in the US. By changing the options above the graph, one can instantly change it into a multitude of different graphs.

Cumulative Provisional COVID-19 Deaths, by Week, in The United States, Reported to CDC

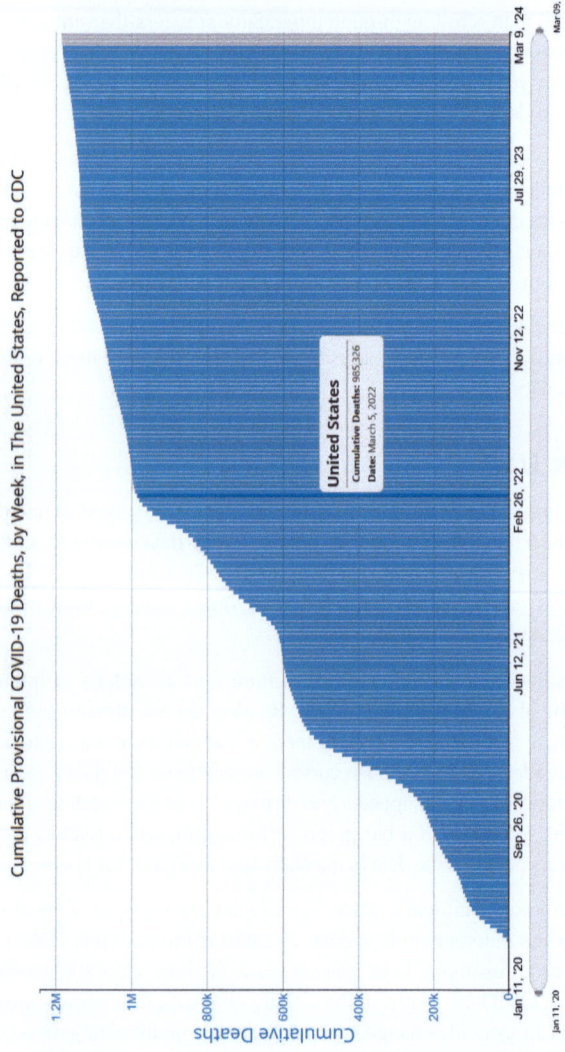

**Figure 6.4** An interactive graph allows easy reading of data.

A "Select a state or territory" drop-down menu allows users to display data for the entire country (which is what is chosen for Figure 6.4) or data for individual US states and territories. A "View (left axis)" drop-down menu allows users to change what the left axis/the bars measure, such as weekly new hospital admissions or weekly deaths. A "View (right axis)" drop-down menu allows users to add a right axis to the graph and display an additional set of data. Multiple choices exist here. Figure 6.5, for example, adds a right axis tracking the new hospital admissions per 100,000 people.

As mentioned in Chapter 5, a risk with popular science graphs is cramming too much data into a single display, which can overload readers' short-term memory. An alternative is to use multiple graphs to parse the data, so each graph is easier to read. Doing so, however, takes up more real estate in print documents. It also limits graphs' ability to juxtapose related data in a single display to tell richer stories.

These dilemmas are more elegantly addressed by interactive graphs. Changed data displays happen within the same "physical" screen space and do not take up additional real estate. More importantly, users have control over how much data and what kind to add. Although multiple $y$-axes significantly add to the visual complexity of a graph (see Chapter 5), users who choose to add an axis are arguably better prepared to process the more complex graph. In addition, because the extra data are added based on a user's choice, as opposed to a designer's pre-determined decision, the added data will come across as more intuitive and meaningful for that user. The real-time change of the graph – selecting an option adds a differently colored line – also allows users to easily see which data bar/line corresponds to which axis.

Not only can users choose how many axes to display, with some interactive graphs, they can also set the scale of the axis: either the arithmetic scale (also known as the linear scale) or the logarithmic scale. As mentioned in Chapter 5, the logarithmic scale is superior for showing exponential changes but is often misunderstood by nonspecialist readers. Displaying both the arithmetic and the logarithmic scale is one potential way to help readers fully appreciate the data. Interactive graphs easily accomplish this task with the click of a button (Figure 6.6).

Interactive graphs can also effectively layer nonquantitative data. As mentioned in Chapter 5, graphs are designed to visualize numbers.

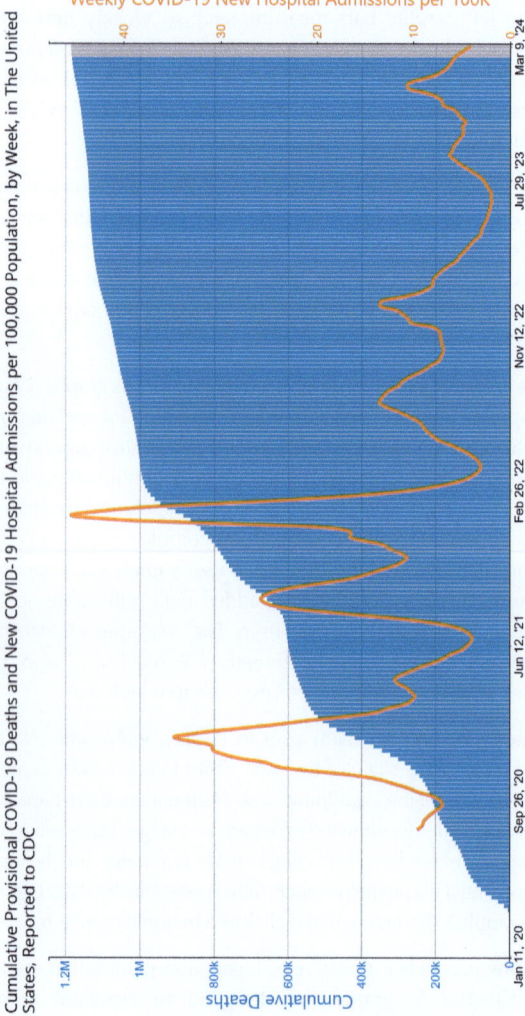

**Figure 6.5** An interactive graph allows the addition of an extra *y*-axis.

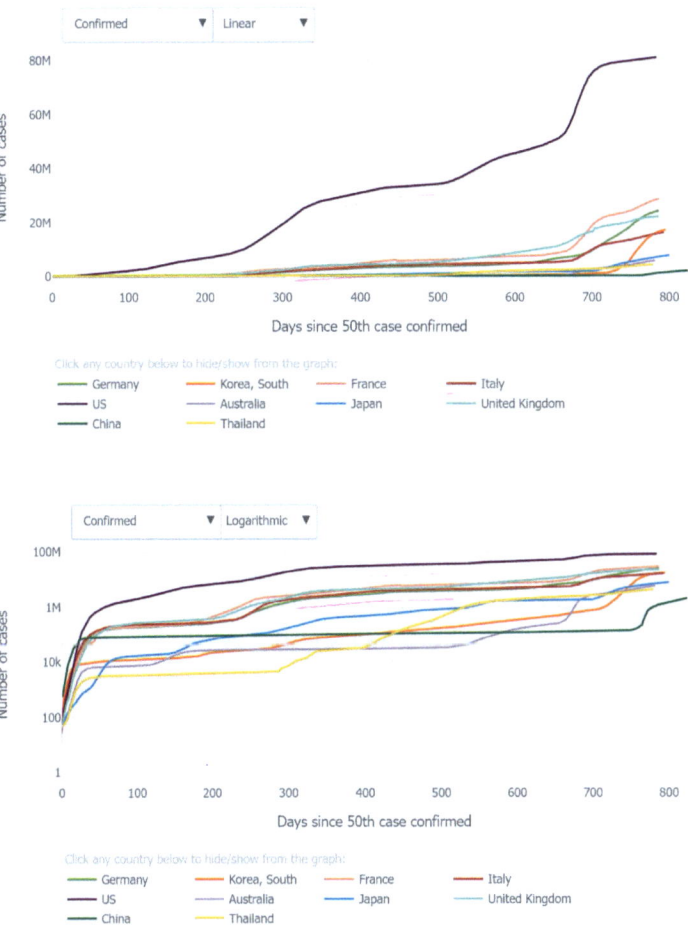

**Figure 6.6**    Toggle between the arithmetic scale and the logarithmic scale.

While we can add text to static graphs, given the limited space of a graph, we risk making the text (and numbers) difficult to read.

With interactive graphs this is less of an issue because data can be selectively read one at a time. Figure 6.7, for example, graphs COVID-19-related state policies alongside the number of new confirmed cases (or the number of new deaths, toggled by the button in the top right corner). The drop-down menu at the top left allows readers to choose states (Illinois is chosen in Figure 6.7).

The solid line in Figure 6.7 records the number of new confirmed cases, as in a typical line graph. The dotted vertical lines indicate key policy events and are color-coded according to their nature. For example, policies that relate to closing down a state (such as limiting gatherings) are in red. Policies that relate to opening up a state (such as reopening state parks) are in green. As readers click on each dotted line, a description of the policy appears at the bottom of the graph. One can also use the "previous" and "next" buttons on the top right to view the policy events in chronological order.

With this setup, Figure 6.7 allows readers to see how the state administration responds to rising and falling case numbers, as well as how different policies may have affected the development of the pandemic. This juxtaposition of quantitative and nonquantitative data provides a layered and rich understanding that would be difficult to realize in static graphs.

The power and versatility of interactive graphs, however, doesn't mean they can gracefully handle any and all data thrown at them. Consider Figure 6.8, which tracks confirmed COVID-19 cases around the globe. At a glance, the graph looks very busy, with dozens of data lines in different colors. On further examination one realizes that the graph provides endless customization. One can use the "Highlight" drop-down menu to select individual countries from a scrolling list (the United Kingdom was selected in Figure 6.8). One can switch between the logarithmic (log) or arithmetic (linear) scales. There is also an "Animate" button that, when pressed, allows the graph's data lines to populate over time.

Other menu options and choices are less meaningful. The "Data" drop-down menu, for example, provides over 40 choices, some of which are shown in Figure 6.9(a). One has to wonder if all of these choices make sense to a nonspecialist audience: What, for example, is $\Delta$ 1 Wk. Avg New Cases/Day?

**Figure 6.7** COVID-19 trends vis-à-vis state policies.

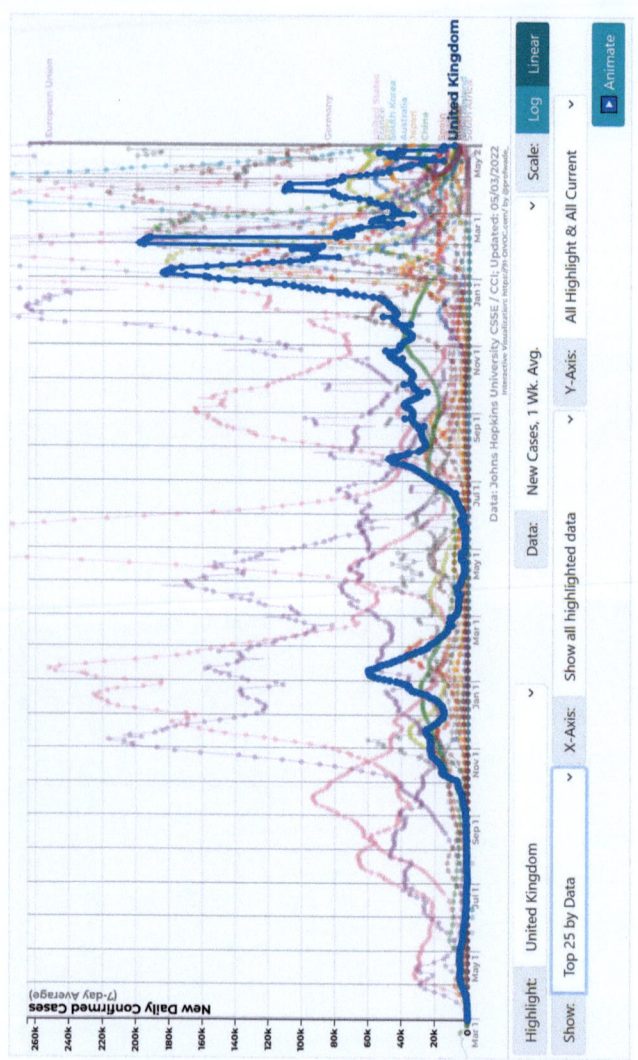

**Figure 6.8** An overly complicated interactive graph.

Moreover, do all of these choices actually result in useful data trends or patterns? Graphs are supposed to be illustrative visual displays, not a repository of any and all data ever collected on a topic.

Still other menu options and choices are plainly confusing. The "X-Axis" and "Y-Axis" drop-down menus contain choices such as "Show all highlighted data," "Show all visible data," or "All Highlight & All Current." It is not self-evident what these wordings mean, and cycling through some of them seems to produce no obvious changes to the graph.

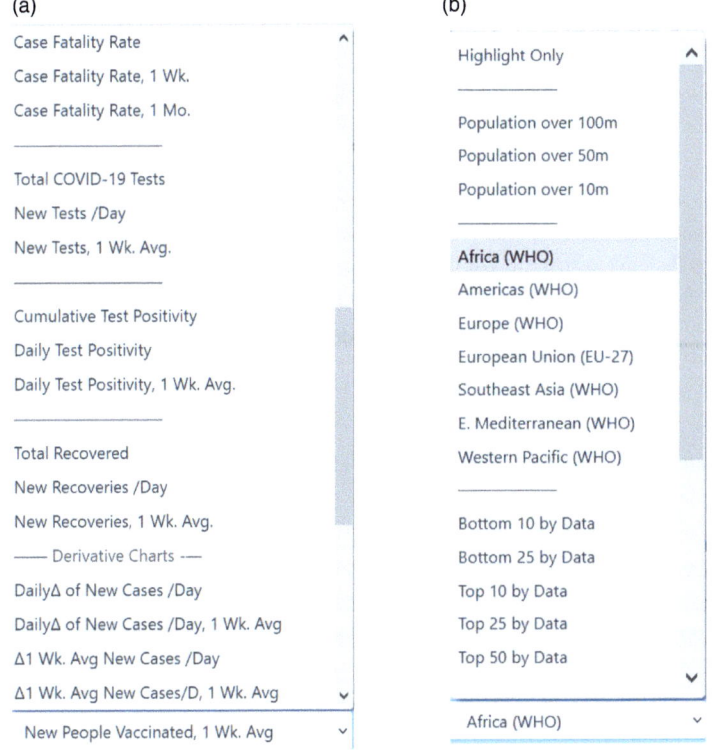

**Figure 6.9** Additional menu options for the graph shown in Figure 6.8.

The "Show" drop-down menu offers about 20 different choices, some of which are shown in Figure 6.9(b). These choices were confusing to me because if I have selected a country (United Kingdom), why am I being asked to show data for "population over 100 m" or for "Africa"? After some experimenting, I realized that these options map additional countries in addition to the selected United Kingdom onto the graph. For example, if Africa is selected, then all the African countries will appear on the graph as well. By default, the "top 25" countries (whatever that means) are shown together with the selected country, which is why the graph contains so many data lines and looks cluttered.

While there is value to situating country-specific data within regional or global contexts, Figure 6.8 violates the basic guidelines of how many data lines to present in a single line graph. All the colorful lines compete for attention, and they are so crowded that there is actually no way to tell them apart.

Without doubt, graphs like Figure 6.9 testify to the technical prowess of interactive graphs to filter, display, and juxtapose diverse sets of data. This *is* what makes them superior to static graphs at handling big data. That said, there is still a limit to how much data interactive graphs can handle – or rather, there should be a limit to how much data we allow them to handle. Simply because visual technology allows a certain function doesn't mean that the function will be meaningful to a nonspecialist audience. Simply because data exist and can be graphed doesn't mean they should be.

## Games

In the mid-2000s, Luis von Ahn, a computer science professor at Carnegie Mellon University, came up with the concept "games with a purpose" (or GWAP). Put simply, GWAP are online games that crowdsource human players to solve large-scale computational problems. The more people who play, the more brain power can be harnessed, and the more complex problems can be solved. Examples of GWAP include having human players describe images so that these descriptions can be used to tag online images and allow screen reading by the visually impaired, or having human players describe songs to allow computer searching beyond using song titles.

In the life sciences, GWAP have been developed to harness humans' visual processing and pattern recognition ability to solve problems that computers struggle with. A well-known example is Phylo.

Phylo was developed in 2010 by McGill University School of Computer Science and Centre for Bioinformatics. It is often dubbed a game that helps crack the genetic code. More specifically, what the game tackles is the task of multiple sequence alignment: To align DNA sequences from different species so researchers can examine how sequences change during evolution, infer the functions of sequence regions, and examine their roles in genetic disorders. Conventionally, this task is performed by computer algorithms, which are, however, prone to inaccuracies because of the large size of the sequences (there are up to billions of DNA bases in mammalian genomes). Phylo developers hoped to harness humans' superior visual recognition ability to improve these alignments.

In the game, DNA bases (A, T, C, and G) are displayed as colorful tiles in multiple sequences (Figure 6.10). Players slide the tiles to match as many same-colored tiles as they can while leaving minimal gaps between tiles.

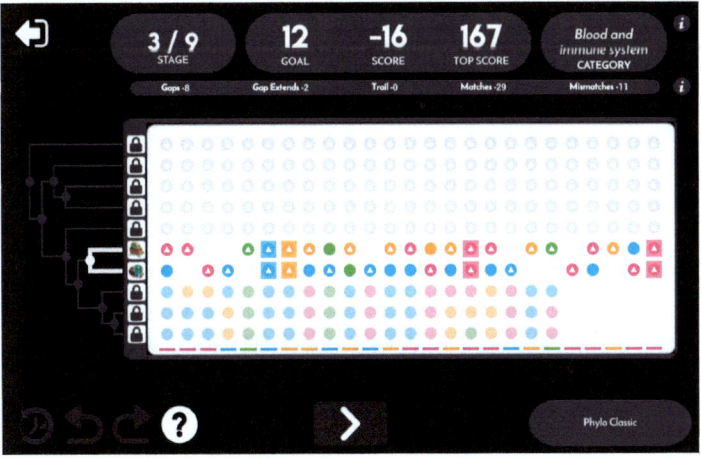

**Figure 6.10**    The Phylo game interface.

Performance is scored based on the matches, mismatches, and gaps. Once players tie or beat an existing score achieved by the computer algorithm, they are deemed to have solved a puzzle and can unlock the next level, where more sequences are added.

Within seven months of its launch, Phylo had attracted 12,000 registered users and just as many unregistered users who together submitted 250,000 solutions. Phylo players made genuine contributions to sequence alignment and came up with solutions that significantly improved those reached by computer algorithms. For this reason, Phylo is hailed as a great success.

And it is, indeed, a great success for science: Important scientific problems were being tackled and solutions were being improved. Yet, the question I'm interested in exploring here is whether the game actually allowed players to understand the genetic code they were supposedly cracking, or whether the game more generally engaged players with science – beyond using their free labor, that is.

According to Phylo developers, education (about computational biology, evolutionary biology, and genetics) *should* be a fundamental aspect of the game. However, judging by the game design, they did very little to facilitate that goal beyond providing some background information via an "About" menu. In fact, to not intimidate players, Phylo developers purposefully hid the science by removing any and all traces of DNA from the game and providing no explanation about the sequences in question other than putting them into categories such as "blood and immune system." This way, "people won't think about the biological problem, but just have fun and be entertained," said Jerome Waldispuhl, a Phylo project leader, to *WIRED* magazine reporter Lisa Grossman. Born out of this design principle, Phylo looks and plays like a Bejeweled- or Tetris-style game, and someone can become an expert player without gaining much, if any, insight into genetics.

It is also worth noting that despite its nonscience design, 42% of registered Phylo players failed to complete a single alignment puzzle, and the majority of the solutions, as Kawrykow et al. reported, came from 10% of the most prolific players. In other words, the public reach of the game is limited.

Phylo is not an isolated example. An even more famous GWAP in the life sciences is Foldit, which was launched in 2008 and co-developed by multiple universities such as University of Washington and University of Massachusetts, Dartmouth.

The purpose of the game, in a nutshell, is to predict protein structures. Proteins are large molecules that play critical roles in our bodies. They give structure to cells, support various cellular functions, and enable the proper working of our tissues and organs. Each protein is made up of amino acids, which are small molecules. Like tiny beads, amino acids are strung together into a long chain to create a protein. A small protein can have 100 amino acids, while a large protein can include 1,000 amino acids. Our body uses 20 common amino acids, strung together in different ways, to produce the numerous proteins we need.

This amino acid protein chain doesn't stay as a stretched, straight line; instead, it folds into a unique, complex structure, known as a protein's native structure. The native structure has the most favorable chemical interactions and the lowest free energy. Cracking proteins' native structures allows us to understand proteins' functions, to intervene when they misbehave and cause diseases, and to explore additional functions that proteins may offer.

But that is easier said than done. Even a small protein can fold into many, many different possible structures. Computer algorithms exist to predict the best protein structures, but this work is computationally expensive as computers must conduct an extremely large amount of random experimentation. It was thought that humans' visual perception and intuition would help address this challenge – and so the Foldit game was born.

Within the game, scientists post proteins with unknown native structures (Figure 6.11). Players use their computer mouses to drag the amino acid chain around while trying to meet certain folding rules: For example, atoms should not be so close to each other that they clash; at the same time, they should not be so far away from each other that there is empty space inside a protein. The game also comes with automatic tools that can help improve the structure with the click of a button (Figure 6.11 bottom). The "shake" tool, for example, can automatically rotate amino acids' side chains in different combinations to avoid clashes. When players think the tool has done a good enough job, they can stop the function. As a player improves a protein's structure, their score increases. A leaderboard showcases the individual/

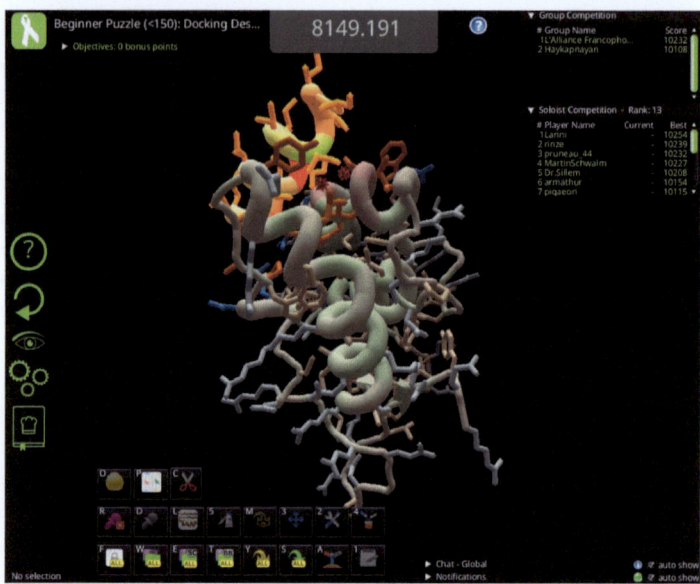

**Figure 6.11**  The Foldit game interface.

group players with the highest scores (Figure 6.11 top right). This creates competition between players to come up with the best structure that may resemble a protein's native structure.

As with Phylo, Foldit players have made genuine contributions to science, coming up with solutions that were comparable to or better than those reached by computer algorithms. One significant example, as Khatib et al. described, concerns the Mason–Pfizer monkey virus (M-PMV) retroviral protease. M-PMV is a virus that causes AIDS in monkeys; its retroviral protease is a protein critical for viral proliferation and thus the focus of drug development. Previous attempts to determine the protein's structure have all failed. Remarkably, when this protein was posed as a puzzle in Foldit, players came up with several solutions that were good enough to allow a rapid determination of the final structure by scientists.

Once again, this is a tremendous success for science. But, once again, is that all there is to games with a purpose? With their advanced interactive visual displays and functions, with the amount of resources and funding (including public funding) that were put into these games, couldn't or shouldn't they also contribute to public understanding of and engagement with science?

Apparently, Foldit developers didn't think so. According to Cooper, Treuille, et al., Foldit as originally designed was not meant to educate players; its sole purpose was to enable nonexpert players to advance science. To do so, these developers, as Phylo developers, tried to make the game "look inviting and fun, and not bring back memories of high school textbooks" (p. 43). The visual displays used in the game do reflect the nature of the scientific problem – but only to the extent that players can make meaningful contributions. The actual complexity of the science is hidden, so the game can attract the maximum number of everyday players.

I should add that during the COVID-19 pandemic, when the need for remote learning surged, Foldit developers added an Education Mode to the game, with the goal of helping teachers teach a protein biochemistry class. The Education Mode provides basic biochemistry knowledge and requires students to memorize amino acid structures. While this mode may help students pass a class, with its emphasis on curriculum-based learning, the general public may not choose to play in this mode.

At the very least, we don't know *what* everyday players of the game are interested in. Existent studies of Foldit mostly concern themselves with how the game is designed and how its design facilitated gameplay and scientific research. There is very little information concerning who the players are, what they do, or what they think of the game.

Only one *Nature* article by Cooper, Khatib, et al. talked about, almost in passing in its supplementary materials, conducting user surveys on the Foldit website. A demographics survey received a total of 149 responses, which is a very small number compared to the (then) 57,000 registered players. According to the survey, the players had a variety of occupations (see Figure 6.12(a)). Incidentally, this graph reflects just how *unimportant* this kind of player research is to the scientific community – for a study published in the prestigious *Nature*, we get a pie graph with no actual numbers either in the graph itself or elsewhere

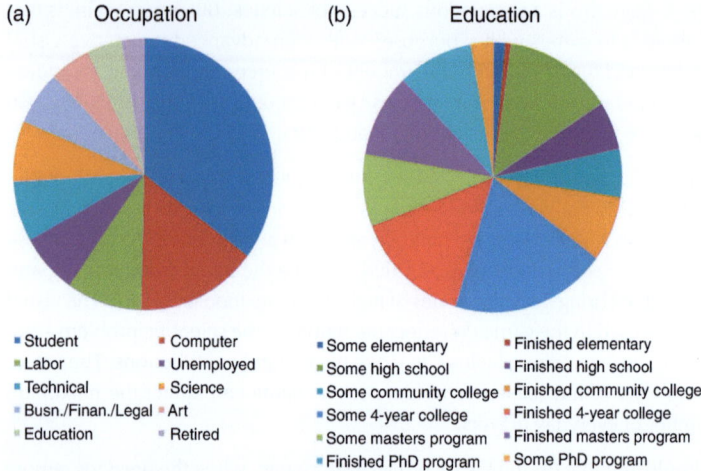

(a) Occupation

- Student
- Labor
- Technical
- Busn./Finan./Legal
- Education
- Computer
- Unemployed
- Science
- Art
- Retired

(b) Education

- Some elementary
- Some high school
- Some community college
- Some 4-year college
- Some masters program
- Finished PhD program
- Finished elementary
- Finished high school
- Finished community college
- Finished 4-year college
- Finished masters program
- Some PhD program

**Figure 6.12**   Demographic survey of Foldit players.

in the text. To the best of my ability, I estimate that "Students" accounted for about 35% of the respondents, "Computer" accounted for about 15%, and the next five occupations ("Labor," "Unemployed," "Technical," "Science," and "Business/Financial/Legal") each accounted for about 7–8%.

The respondents' education backgrounds also varied considerably (Figure 6.12(b)). This pie graph likewise provides no actual numbers. Worse still, some of the slice colors are very similar, and some slices (the PhD program ones) seem out of order. To the best of my ability, I estimate that "Some 4-year college" is the largest slice, accounting for about 20% of the respondents, followed by "Some high school" and "Finished 4-year college."

Between the two graphs, it seems that students, as opposed to the general public, were a significant portion of the players. This makes sense because biology and biochemistry teachers have been using Foldit as a teaching aid. Moreover, as with Phylo, the number of active players is low: while there may be 57,000 people who ever registered, there were only about 70 distinct top players. So, once again, the public reach of the game is limited.

Cooper, Khatib, et al. also surveyed players on why they play the game. This survey had even fewer responses (48), and the results were again reported as a numberless pie graph. About 40% of the respondents played the game for science-related purposes (e.g., "To crack the protein folding code for science" or "To understand the folding process better"). About 35% played it for fun (e.g., "It's fun and relaxing"). About 20% played for achievement (e.g., "To get a higher score than the next player"). The rest, a very small percentage, played for social connection. If this small survey is representative, then over half of the Foldit players did not relate their gaming experience to science. There was, in other words, no engagement with science.

Games with a purpose, like Phylo and Foldit, fall under the umbrella of citizen science projects, which are projects that employ everyday citizens to collect, categorize, or analyze scientific big data. Citizen science as originally conceived easily emphasizes the one-way contribution of citizens to science (e.g., employing bird watchers to gather bird population data). But scholars increasingly argue that these projects should not merely use participants as research tools, but should allow them to gain relevant insights and engage them in addressing questions regarding their personal or local concerns.

By this, I do not mean that games like Phylo and Foldit should be packed with didactic lectures. Certain accessible, relevant scientific knowledge will be useful to add to the game interface, but more importantly the game developers need to put more emphasis on studying the game players and their needs rather than obsessing over how to effectively use players as research tools. From there, we can hope to layer more engaging scientific contexts, more meaningful activities, and more scientist–public interactions into the games. The work will be exploratory, and the results may be initially messy, but that is a step in the right direction to realizing genuine engagement.

## Three-Dimensional Models

Molecules, as we saw above, are three-dimensional objects. Understanding the three-dimensional structures of molecules, especially large molecules like DNA and proteins, is essential to understanding life. When James Watson (1928–) and Francis Crick (1916–2004), with the help of Maurice Wilkins

(1916–2004) and Rosalind Franklin (1920–1958), solved the double-helix structure of DNA, they revolutionized life science research.

Watson and Crick's model was built with cardboard cut-outs, metal plates, and rods, completely outdated by today's standard. Today, we have sophisticated computer programs that can model molecules in dazzling colors, high resolutions, and various styles. Users can use balls to represent atoms and sticks to represent the bonds between atoms. They can use smooth surface or mesh to represent the overall shapes of molecules. They can adjust angles to see a molecule from different perspectives. They can magnify selected elements for a close-up examination.

These computer models are deemed a revolutionary change in the research and education of biochemistry and molecular biology. Studies show that their convenient and versatile visual displays help researchers to explore and students to understand the structures of molecules. Accordingly, current research on molecular visualization focuses on how to develop ever more sophisticated hardware, software, and systems that meet the needs of scientists and scientists-in-training.

There is also the assumption that these models, as Hirst et al. put it, will "engage the public and thereby increase public awareness and understanding of scientific problems" (p. 15). But there is no actual evidence to back up this feel-good sentiment. We do not have research on the needs of public audiences or how computer modeling can be used to bridge those needs and formal sciences. In addition, many benefits of three-dimensional models derive from the fact that in research and formal education settings, a user has access to visualization programs and can directly manipulate a model. Most public readers do not have this access.

Given this reality, newer, faster, and more complex tools are less likely to benefit the public than thoughtful adaptations of the tools we already have. Notably, internet- and server-based software allow us to run increasingly complex applications without having to obtain expensive programs, install them on computers, or figure out complex settings.

Jmol, for example, is an open-source viewer of molecular structures. Users can access Jmol using an internet browser to examine and interact with molecular

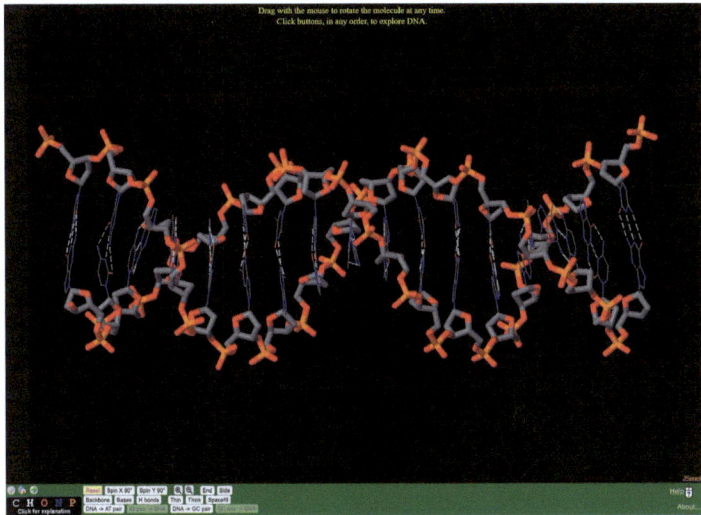

**Figure 6.13**    A Jmol tutorial for the DNA molecule.

models; they can change model styles, rotate, zoom in and out, and hide or show selected elements. Jmol tutorials have been developed that allow novice users to explore common molecules such as DNA (Figure 6.13).

These efforts, however, are still underdeveloped and haphazard. There aren't that many public-facing modeling tutorials to start with, and quality varies for the ones that we do have. To take full advantage of interactive web applications for public awareness and engagement, we need not only efforts to develop more tutorials, but research on what public users actually desire from these tutorials: What do they want to see, how, and why? Only then can we hope to realize the assumption that modern visualization technologies that are developed for specialist audiences will engage the public and increase their awareness and understanding of science.

Because molecular modeling involves proprietary software, to some extent it is understandable that public needs fall by the wayside. Yet, unfortunately, we see the same happen with the so-called "publicly available" data. I am referring to visual data that are collected and curated by central governments, such as the US federal government. These data frequently fall into the public domain, where they can be freely used by anyone. Yet, all this talk of "public" doesn't mean that the visual databases actually serve the needs and interests of everyday citizens. Let's, for example, look at the Visible Human Project developed by the US National Library of Medicine.

The Visible Human Project consists of anatomically detailed, three-dimensional representations of a male human body and a female human body. The project scanned two cadavers from head to toe, at millimeter intervals, using computed tomography (CT) and magnetic resonance imaging (MRI). Then, the cadavers were frozen, milled into millimeter-thin layers, and digitally photographed one layer at a time, resulting in over 7,000 anatomical images. The Visible Man data set was publicly released in 1994, and the Visible Woman followed in 1995.

Although anyone with an internet device can navigate to the National Library of Medicine website, find the Visible Human Project, and view the myriad visuals, these visuals would make very little sense as isolated, static, cross-sectional photographs, CT scans, and MRI scans (Figure 6.14).

What are we looking at here? (It is a cross-sectional image of the female legs.) What is the significance of this visual? (Not much on its own for an everyday viewer.) What can we do with it? (Nothing really.)

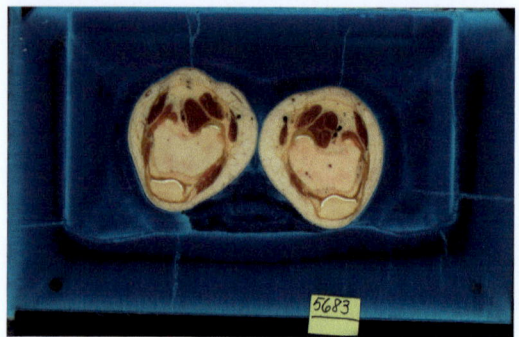

**Figure 6.14**   A sample image from the Visible Human Project.

What we need are interactive models of these raw (no pun intended) visuals. You see, the cross-sectional visuals can be stacked back up to reformulate the body in various ways: Tissues can be simulated, bones can be reconstructed, and the inside of organs can be probed. Theoretically, this database can lead to intriguing interactive models that allow the public to understand and explore the human body in endless ways. For example, as Daston and Galison wrote, viewers can examine sections of the body, rotate the sections, highlight certain elements, remove other elements, or produce two-dimensional images from any angle.

Some of this work *was* done. Museum exhibits were created where holograms show the muscles, skeletons, and tissues of the body from different angles. Animated videos exist on YouTube that offer a "fly-through" of the human body. Yet, by and large, these so-called publicly available data serve the needs of scientists and scientists-in-training. For example, mannequins were developed to help surgeons practice surgical procedures. Anatomy lessons were built for medical students to use during lectures, labs, or self-study. By and large, these resources require institutional subscriptions or are physically inaccessible to the public and, without research evidence, are unlikely to serve the needs or interests of public viewers.

## Conclusion

Interactive visuals are the future of big data life science communication. When they are published online and made interactive, visuals take on a multitude of advantages: They allow instant access, customized filtering, and user-driven actions. Objects of interest can be rendered in three dimensions and examined in multiple ways to inspire interest and enhance understanding.

Still, there is an upper limit to how much we can pack into an interactive visual. At a certain point, the sheer number of choices and amount of information will become visually and conceptually overwhelming.

More importantly, we need to admit that the advance of computer visualization technologies alone does not equate to meaningful public understanding of and engagement in science. Current technologies – and studies about those technologies – focus heavily on formal science education and research. While valuable in their own ways, these efforts ignore the needs of the public or are only interested in harnessing their labor. To create interactive visuals that are truly useful to the public, we need to devote serious attention to researching public needs and developing platforms that cater to those needs.

# 7 Infographics

As we have seen throughout this book, standalone visuals like photographs and illustrations are promising ways to communicate science to the public – and they carry their fair share of misconceptions and complications. These promises – as well as challenges – are multiplied in infographics.

The word "infographic" comes from the phrase "information graphic." Originally, the term referred to the production of graphics for print media such as newspapers and magazines. Today it refers to a unique multimodal genre that combines data visualizations (i.e., graphs such as lines, pies, bars, and pictographs), illustrations (such as icons and drawings), photographs, and small amounts of text. When designed for online use, infographics can also have interactive components. For example, putting the mouse cursor somewhere on the infographic may reveal a small pop-up window with additional information. Some infographics are also animated: bars in a bar chart may grow, colors may change, or characters may move. This is often achieved by using animated GIF files that display a sequence of static images in a repeating loop, which creates the illusion of motion.

Infographics are considered an effective way to communicate complex information to a large audience. The use of fun and colorful visuals can attract viewer attention and enhance information recall. Text and visuals can complement each other, allowing viewers to form both a verbal and visual mental model to enhance understanding. A short one- or two-page infographic will appear far less intimidating to read than a lengthy report. Given these advantages, business corporations, research institutions, government agencies, and media outlets are all designing and publishing infographics, in topics ranging from home improvement to computer technology to higher education.

In the life sciences, infographics are used to illustrate the human body, plant lives, all kinds of nonhuman animals, and more. One particular venue where infographics have proliferated – for good reasons – is public health communication. Public health communication frequently features complex scientific research and quantitative data; it needs to reach diverse audiences, including nonnative speakers of a given language; it needs to attract and engage audiences for them to make positive health changes. Infographics are well positioned to meet these goals – at the same time, they are also susceptible to various misconceptions, as we will see in this chapter.

The infographic examples featured in this chapter focus on public health communication, both because this is an essential area for infographic use and because these infographics are often copyright-free. The lessons learned from these infographics apply to other life science infographics. Because the unique challenges with interactive visuals (some of which can easily be called interactive infographics) are already covered in Chapter 6, this chapter focuses on static infographics.

## Some Design Frameworks

Given the combined and complex use of visual and textual elements in infographics, there is a very long list of applicable design principles. It is not the purpose of this chapter to delineate them all; indeed, it is not possible for a single chapter to delineate them all. What follows are brief introductions to several comprehensive design frameworks; interested readers can find out more about them by following the cited sources. Design principles relevant to our misconceptions will be further elaborated on later.

Baxter et al., based on a review of the literature, proposed a total of 84 guidelines for designing infographics. These guidelines are organized into seven categories: cognitive principles, typeface principles, Gestalt principles, hierarchy/structure principles, color principles, illustration principles, and graph principles. For example, within the category of cognitive principles, designers are advised to highlight the most important information so as to optimize viewers' learning, to avoid displaying too much information lest they cause viewer cognitive overload, and to create visuals that are familiar to readers so as to facilitate comprehension. Within the category of typeface principles, designers are suggested to limit the number of typeface families to two or three in a single infographic, to choose simple typefaces that are easy to read, and to justify text to the left to enhance legibility.

The Gestalt principles, in particular, deserve a brief explanation as some readers may not be familiar with them. Gestalt principles were established in the early twentieth century by a group of German psychologists. The German word "gestalt," which means shape, is roughly interpreted to mean "pattern." Thus, Gestalt principles explore how the human visual system perceives patterns out of visual stimuli. Baxter listed nine Gestalt principles.

The Gestalt principle of proximity, for example, states that things that are physically close to each other will appear more related than things that are further apart, regardless of their actual relationships. In an infographic, if a graph and a short passage of text are placed closer to each other than they are to other visual or textual elements, readers will automatically assume that the text explains the graph. Designers, then, need to pay attention to where they place elements on an infographic lest they inadvertently cue relationships that do not exist or fail to cue relationships that do exist.

The Gestalt principle of figure–ground states that the human brain perceives elements on a page either as unimportant background (ground) or objects of focus (figure). In an infographic, designers can use strong contrast to help viewers separate the figure from the background. For example, warm colors such as reds and yellows can be used to highlight figures on a background of cool colors such as blues and greens.

Stones and Gent proposed seven categories of infographic design principles: know your audience, restrict color, align elements, prioritize parts, highlight the heading, invest in imagery wisely, and choose charts carefully. The category of restricting color, for example, suggests that designers use color to enhance focus and appeal but restrict color palettes to 3–5 colors and to check color for legibility. The category of prioritizing parts suggests that designers should not treat all parts of an infographic equally, but should emphasize important parts by using things such as larger fonts.

Lonsdale and Lonsdale proposed five categories of design guidelines for info-graphics: general guidelines, text/typography, color, graphics/visual elements, and layout and structure. Each category has anywhere between 16 and 39 principles. Under general guidelines, for example, designers are urged to form a clear focus and purpose for their infographics, to design infographics for quick and clear communication, and to limit infographics to under two pages. In the layout and structure category, designers are advised to use consistent typefaces, shapes, colors, alignment, and spacing; to create a balanced placement of visuals and texts; and to use a grid to organize content on the page.

## Simply Using Visuals Doesn't Good Infographics Make

Everything being equal, the more "visual" an infographic looks, the more likely it is to catch readers' attention, speed up communication, and make its key message memorable. By contrast, a wall of text is likely to intimidate readers. Indeed, as Krum wrote, when viewing an infographic, readers tend to focus on the visual elements and skip the text, considering the latter unimport-ant and/or too time-consuming to process.

These common beliefs and habits make visuals the signature feature of info-graphics, as they should be. At the same time, they also fuel the misconception that as long as we put multiple visuals into a design, we can turn a document that may otherwise be called a "handout" into an infographic. It doesn't help that we can *always* find somewhat relevant visuals to use no matter the subject we are communicating.

However, readers do not just want to see visuals of superficial relevance, like a picture of cars when fuel consumption is the topic of discussion. Such visuals are merely illustrative of the text and do not add intrinsic value. Instead, as Polman and Gebre's study shows, readers want to see visuals that embed data and convey new information. These are the visuals that can help support the key message of an infographic.

Consider Figure 7.1, which is a so-called infographic from the US Department of Agriculture's MyPlate initiative, which is designed to promote healthy eating. The intent of Figure 7.1 is obvious: to demonstrate the five dishes people can make to consume brown rice, a whole-grain rich in vitamins,

**Figure 7.1** An infographic, or, rather, an illustrated bullet list.

minerals, and dietary fiber. For each dish, Figure 7.1 gives its name, posts a picture, and then lists the ingredients in bulleted items. Photographs of brown rice also line the top and bottom of the document. If we remove all the visuals in Figure 7.1, the document still makes perfect sense, loses very little (if any) essential information, and becomes a simple bullet list. And that is what it really is: an illustrated bullet list. To be a true infographic, we need visuals that add value and explain, for example, the nutritional profile or health benefits of brown rice.

Unfortunately, examples like Figure 7.1 are everywhere. In some of these examples we see the use of illustrative icons as opposed to photographs. Icons are a favorite in science infographics because they are minimalist and appear more formal and professional. Large icon databases also guarantee that we can pretty much find icons for any life science subject matter: test tubes, microscopes, bacteria, DNA, scientists, etc. Because of this, it is easy to write up some texts, put icons by the side, deposit them into a premade template, and call it an infographic, which is what Figure 7.2 did.

Figure 7.2 shows part of an infographic on antibiotic resistance (the entire infographic is long and narrow and difficult to reproduce in its entirety). The excerpt shown here reflects the infographic's overall style. As can be seen, an icon accompanies each text snippet. But the icons only illustrate the text and are marginally successful at that: colorful pills, for example, do not necessarily mean "use antibiotics wisely." Either way, the result is a glorified bullet list, not an effective infographic.

## Crowding a Page Doesn't Good Infographics Make

Because infographics are, by definition, a combination of diverse visual and textual elements, there can be a misconception that when we fill a page with assorted materials, we will be creating an infographic. Moreover, following the same logic, the more information there is, the better.

In reality, an essential quality of good infographics is having a clear, focused purpose. In other words, an infographic needs to deliver a simple, key message, and all the data, visuals, and text should support that message. Irrelevant or less relevant content only dilutes the key message and makes it difficult for

# HOW CAN WE STOP IT?

**1. Improve labs:**
Countries need medical labs
to identify bacteria and choose
the right drugs to treat them.

**2. Collect and share data:**
Countries need systems to track
cases and report results globally
to make better policy decisions.

**3. Use antibiotics wisely:**
To ensure antibiotics are here
when we need them, they
must be prescribed and taken
correctly now.

**4. Take measures to
prevent infections:**
Especially in
healthcare settings,
good infection control
practices are critical
to stopping spread of
resistant germs.

**Figure 7.2**    Another illustrated bullet list, with icons.

readers to find important data. Moreover, precisely because infographics
feature multiple and diverse elements, we need to be vigilant about preserving
white space, also known as negative space, which is empty space around the
content of a page. White space provides breathing room, so an infographic
doesn't become overwhelming for a viewer to look at or think about.

Figure 7.3, for example, is an infographic where we have too much information. At a glance, it is visually busy with drawings, signs, background images, bright and highly saturated colors, and text blocks. There is very little, if any, white space where one's eyes can rest. Because so much is crammed in, frequently text and visuals are shrunk to tiny sizes that are difficult to read.

What's more, not all of the information is essential or even necessary. For example, the tagline "Follow some simple food safety advice to keep you and your guests feeling festive this winter" is redundant when we already have the title "Your map to a food-safe holiday." Removing the tagline is not going to detract from the infographic. Similarly, warnings to keep food below 40°F or above 140°F are repeated some five or six times. While repetition may seem like a good idea to emphasize important content, it also means we are dividing up valuable space and that each instance of repetition can only occupy a small space and becomes less noticeable. If we highlight important content only once, we can devote more physical space to it, which will result in larger visuals and text that stand out to viewers as they skim an infographic.

Worse still, multiple visuals in Figure 7.3, such as the yield sign, the do-not-enter sign, and the colored backdrop with snowflakes, contribute no information. Arguably, these visual decorations, by staying true to the map metaphor and the winter theme, can attract viewer attention. But it is far preferable to make design elements do the double duty of attracting viewers and conveying real information, as Figure 7.4 does.

Comparing Figures 7.3 and 7.4, one can see the difference between a cluttered and a clean design. Figure 7.4 (also an excerpt from a long infographic) features ample negative space, which taps into the figure–ground Gestalt principle mentioned earlier. That is, viewers immediately perceive the visuals and text as important (figure) and the empty space as unimportant (ground). This helps them to zero-in and pay attention to the actual content of the infographic. To put it another way, even though Figure 7.4 is much more subdued in its use of visuals and colors, its calm background allows its data visualization, icons, and numbers to stand out and attract viewer attention.

Negative space also breaks the infographic into multiple chunks. Cognitive scientists believe that this chunking effect improves our cognitive function. As mentioned in earlier chapters, humans' short-term memory is notoriously

**Figure 7.3** A crowded infographic.

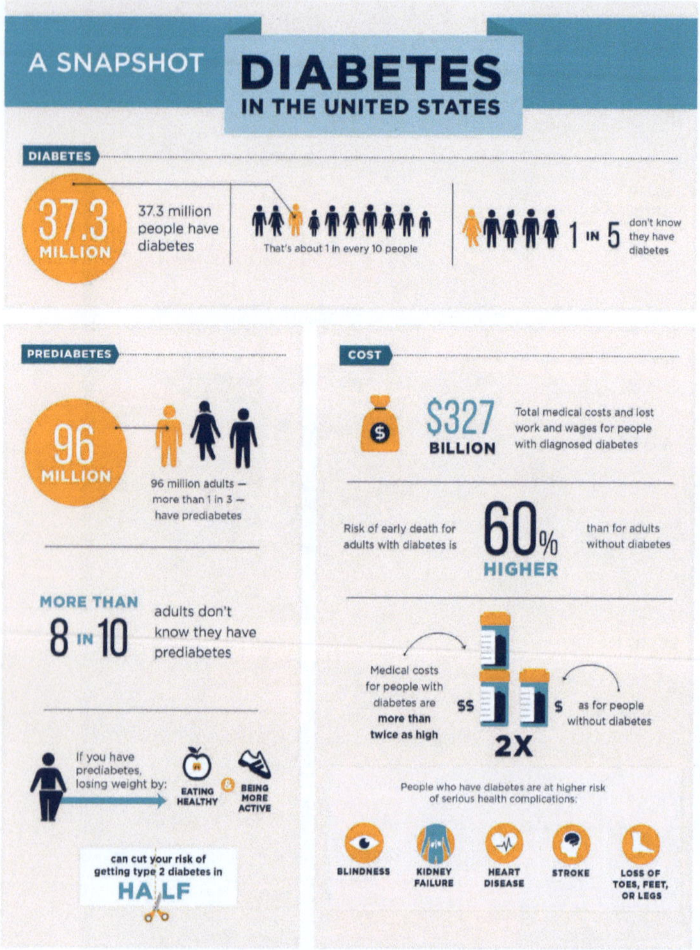

**A SNAPSHOT**

# DIABETES
## IN THE UNITED STATES

**DIABETES**

**37.3 MILLION**

37.3 million people have diabetes

That's about 1 in every 10 people

**1 IN 5** don't know they have diabetes

**PREDIABETES**

**96 MILLION**

96 million adults — more than 1 in 3 — have prediabetes

**MORE THAN 8 IN 10** adults don't know they have prediabetes

If you have prediabetes, losing weight by: **EATING HEALTHY** & **BEING MORE ACTIVE**

can cut your risk of getting type 2 diabetes in **HALF**

**COST**

**$327 BILLION** Total medical costs and lost work and wages for people with diagnosed diabetes

Risk of early death for adults with diabetes is **60% HIGHER** than for adults without diabetes

Medical costs for people with diabetes are **more than twice as high** $$ **2X** $ as for people without diabetes

People who have diabetes are at higher risk of serious health complications:

**BLINDNESS**   **KIDNEY FAILURE**   **HEART DISEASE**   **STROKE**   **LOSS OF TOES, FEET, OR LEGS**

**Figure 7.4**   An infographic with ample negative space and a clean design.

limited. Miller theorized that at a given time we can only "hold" about seven pieces of separate information in our minds. Chunking creates a way around this limit by combining multiple, related pieces of information into one chunk so the overall information we can process increases. For example, rather than trying to remember a telephone number as 10 digits, we remember it as three chunks: xxx-xxx-xxxx.

Visual chunking works the same way. Figure 7.4 provides, minimally, 10 separate pieces of data/information. Yet, visually, it features only four large chunks: the title banner, the visual chunk immediately below titled "diabetes," the lower left chunk titled "prediabetes," and the lower right chunk titled "cost." (Some viewers may see the title and the "diabetes" section as one chunk.) This visual chunking allows Figure 7.4 to package more information without overwhelming viewers the way Figure 7.3 did.

## A Lack of Stories Doesn't Good Infographics Make

With their combined use of visuals and text, infographics are superb at telling stories. As scholars in anthropology, psychology, literature, and other fields show, stories are a primary means by which people make sense of experience, build knowledge, and inform or persuade others. Tapping into storytelling allows infographics to truly demonstrate their communicative value.

Of course, by "stories," I do not mean "fictions." Rather, I mean intelligent ways to make sense of information and data – for example, a story of cause and effect, trends and patterns, problems and solutions, etc.

Given these diverse possibilities, what are the common elements that constitute a story that designers can look for? As Edwards wrote, there are four basic elements: character, setting, narrator, and plot. The first three elements are usually easy to satisfy. The particular subject of a given infographic (whether it be people or inanimate entities) can be characters. Setting is the context where the information in question is applied. The narrator can usually be assumed to be the person who created the infographic. The one element that is often difficult to satisfy is the plot, which is the logic developed through successive events or actions.

Consider Figure 7.5, which features the tomato and how it satisfies the daily veggie consumption requirement for a healthy diet. Tomato is the obvious character here, and it is effectively emphasized via a large illustration and text. Setting is also clear: that of having healthy meals at home, in school, etc. The plate visual and text help to emphasize this setting. The narrator is the designer of the infographic and presumably a specialist in nutritional health.

But there is no apparent plot to it. There is no progression of events, certainly no surprising turn of events, no cause and effect. In fact, the different parts of the infographic seem loosely connected: Why would the "*vary* your veggies" heading be illustrated with a plate, for example, and why would a court ruling be mixed up with ways to eat tomatoes?

By contrast, Figure 7.6 is an infographic with more successful storytelling. Figure 7.6 explains the concept of contact tracing. Here, the characters are Ebola patients and potential patients, which are depicted using icons and text. The setting is an Ebola outbreak, as the subtitle and body text explain. The narrator is the infographic designer and presumably a healthcare specialist. The plot unfolds through the step-by-step sequential and cyclical processes one follows in order to execute contact tracing.

Are some subject matters simply more conducive to storytelling than others? That is possible. Still, I argue that with careful planning we can almost always come up with meaningful plots – doing that work is the very challenge of creating effective infographics. With the tomato infographic example, one potential plot that is practically jumping off the page is the whole question of whether tomato is a fruit or a vegetable. This is a fun debate and even has historical backstories, as Figure 7.5 explains. Leaning into it can potentially make a fun and memorable infographic.

## A Haphazard Organization Doesn't Good Infographics Make

Because an infographic houses an array of text and visuals, which seem to just be scattered on a page, some may assume that an infographic doesn't need to have a rigid organization the way, say, a research article does.

That is not true. Although pieces of information aren't necessarily laid out sequentially one after another in an infographic, readers do intuitively follow

# TOMATO

## VARY YOUR VEGGIES

Aim to make half your plate fruits and vegetables. Tomatoes are a nutritious addition to help you get there.

## HOW IT FITS INTO MYPLATE

A 2,000 calorie diet has a daily Vegetable Group target of 2½ cups. By eating 1 large tomato, you're almost halfway there!

   =

2½ CUPS    **100%** VEGGIE GROUP TARGET

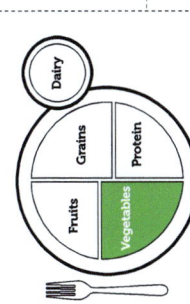 =

1 LARGE TOMATO (1 CUP)    **40%** VEGGIE GROUP TARGET

To find your food group targets, go to ChooseMyPlate.gov/Checklist

## WHAT IS IT?

One of America's most popular garden veggies, the tomato comes in hundreds of varieties. Enjoy fresh in the summer and canned year-round.

## FUN FACTS & TIPS

 Though tomatoes are botanically a fruit, the Supreme Court ruled them a vegetable in 1893.

 Add slices of tomato to your sandwich as an easy way to work toward your Vegetable Group target.

 Tomatoes are versatile! Mix them with melon for a fresh summer salad.

MyPlate
MyWins    For more information go to ChooseMyPlate.gov
USDA is an equal opportunity provider, employer, and lender.

August 2017    USDA

**Figure 7.5** An infographic without stories.

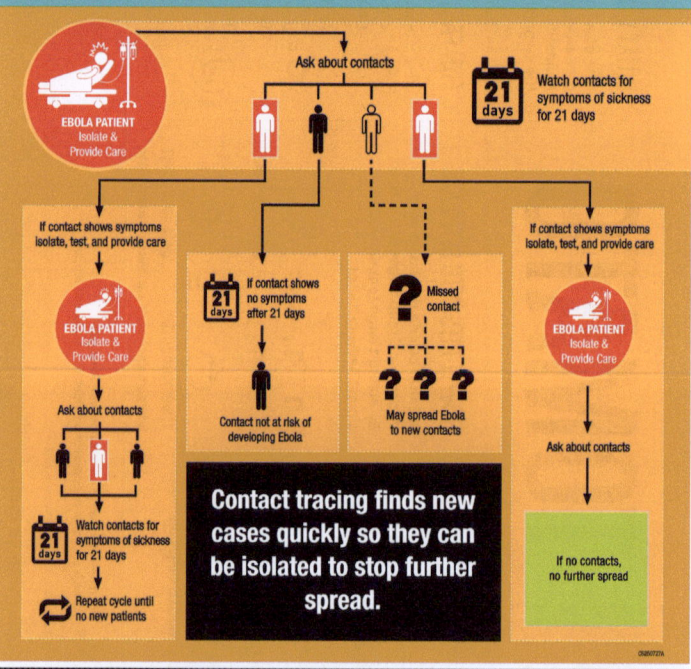

**What is contact tracing?**
Contact tracing can stop the
Ebola outbreak in its tracks

U.S. Department of
Health and Human Services
Centers for Disease
Control and Prevention

**Contact tracing** is finding everyone who comes in direct contact with a sick Ebola patient. Contacts are watched for signs of illness for 21 days from the last day they came in contact with the Ebola patient. If the contact develops a fever or other Ebola symptoms, they are immediately isolated, tested, provided care, and the cycle starts again—all of the new patient's contacts are found and watched for 21 days. **Even one missed contact can keep the outbreak going.**

Ask about contacts

**EBOLA PATIENT**
Isolate &
Provide Care

**21 days**   Watch contacts for symptoms of sickness for 21 days

If contact shows symptoms isolate, test, and provide care

**EBOLA PATIENT**
Isolate &
Provide Care

Ask about contacts

**21 days**   Watch contacts for symptoms of sickness for 21 days

Repeat cycle until no new patients

**21 days**   If contact shows no symptoms after 21 days

Contact not at risk of developing Ebola

? Missed contact

? ? ?   May spread Ebola to new contacts

If contact shows symptoms isolate, test, and provide care

**EBOLA PATIENT**
Isolate &
Provide Care

Ask about contacts

If no contacts, no further spread

**Contact tracing finds new cases quickly so they can be isolated to stop further spread.**

**Figure 7.6**   An infographic that tells stories.

certain sequences defined by the reading direction of their languages. In the English language, that is left to right, and top to bottom.

Moreover, the best infographics, as Krum explains, are organized just as an intelligent article or speech is. They open with an introduction to let readers know what they are reading and why. This introduction can be accomplished by the title of the infographic or by a short snippet of text and some visualization of background data. Then, there will be detailed data and some text to support the key message; these details, as shown above, are broken down into multiple chunks. The infographic ends with some sort of revelation, conclusion, or, if appropriate, a call to action so readers know what they should do with the information they just received. This organization is, essentially, the logic that helps to reveal the story told in an infographic.

To ensure a logical organization, designers need to analyze all the pieces of information to be included in an infographic, establish their relative importance, determine their relationship, and then meaningfully decide how to organize them all. For example, part A needs to come before part B, which needs to come before part C, because A–B–C constitutes a step-by-step process. Or, part D must come before part E, because part D presents a problem and part E provides a tentative solution.

When this work isn't carefully done, we end up with an infographic like that shown in Figure 7.7, which advises its American readers to eat healthy, follow the *Dietary Guidelines for Americans*, and make every bite count. At a glance, Figure 7.7 looks professional enough: It has muted colors, unobtrusive visuals, and interesting-looking data visualization (though it appears bogged down with text). But its organization has much to be desired – in fact, with an optimal organization we may very well be able to reduce its text.

Figure 7.7 has three main sections, signaled by the larger texts serving as headings. The designer explicitly labeled the three sections 1, 2, and 3 to cue readers of the intended organization. However, when readers attempt to follow that organization they will quickly realize that it doesn't work.

We begin with "Start with the 4 guidelines," which is somewhat abrupt and vague: What guidelines, start with them to do what, and why? To put it another way, "Start with the 4 guidelines" is a call to action. Yet, the

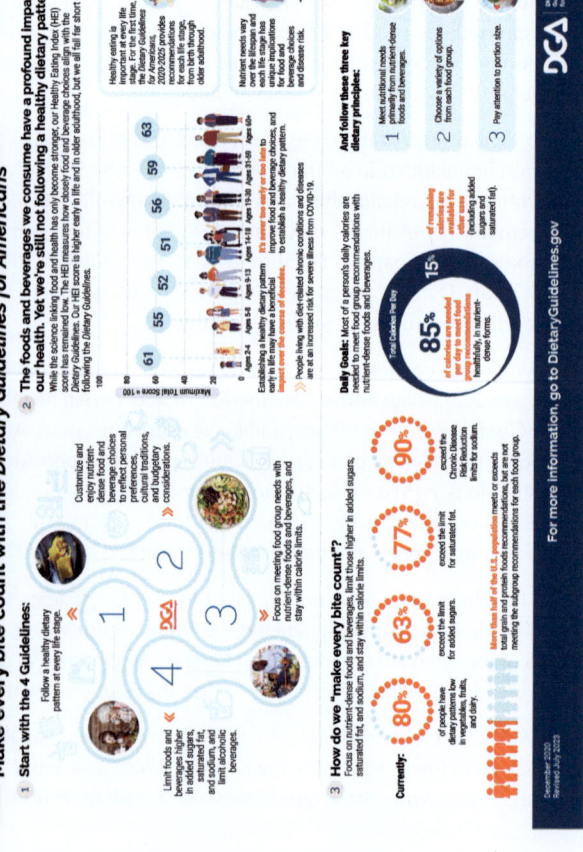

**Figure 7.7** An infographic that lacks coherent organization.

infographic hasn't done the work of establishing a problem, so this call is likely to fall flat.

Section 2 then goes backward to state the problem: "The foods and beverages we consume have a profound impact on our health. Yet we're still not following a healthy dietary pattern." In this section, we also read about the general context of the infographic: the *Dietary Guidelines for Americans, 2020–2025*. All of this is introductory material that should have appeared sooner to create the setting for the infographic.

Finally, section 3 asks "How do we 'make every bite count'?" Well, if we are only now getting to the how-to, then what are the guidelines in section 1 all about? To make matters worse, in this final section we are still looking at data on how poorly Americans do in their diet (e.g., 80% of people have dietary patterns low in vegetables, fruits, and diary). This, clearly, is not *how* to make every bite count.

Figure 7.7's poor organization is partly responsible for its overflow of text. Because the topic goes back and forth, so does the text. For example, the suggestion to focus on nutrient-dense foods and beverages appears six times in multiple sections; the idea that eating healthy is important for every life stage is also repeated multiple times across sections.

By contrast, when an infographic is well organized, information can be streamlined and readers will automatically know where their eyes should go – and the information they pick up along the way will make logical sense. In Figure 7.8, an infographic about the fight against global tuberculosis (TB), the top-to-bottom reading sequence conveys the underlying relationship between the infographic's chunks: There is a problem (TB is globally prevalent), there is hope (TB can be successfully prevented and treated), and the CDC is contributing to that effort.

## Visual Appeal Alone Doesn't Good Infographics Make

As all standalone visuals we have seen in this book, infographics that *look* appealing will catch more viewer attention and be more memorable. More importantly, when people deem an infographic visually pleasing, they are

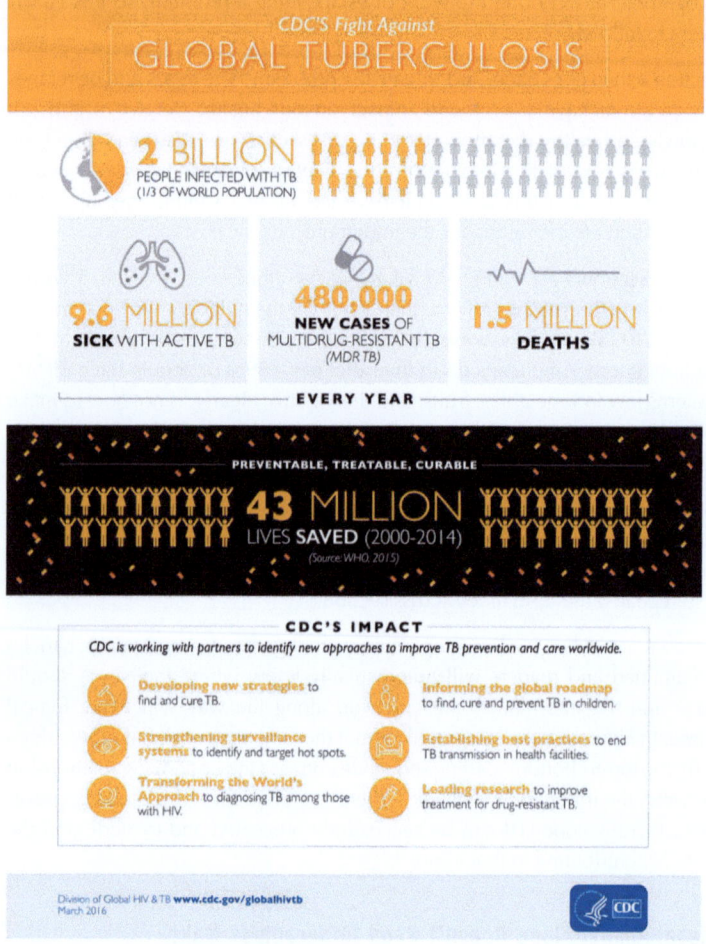

**Figure 7.8**   A well-organized infographic.

more likely to spend time examining it to figure out the story behind it. In other words, visual appeal facilitates audience engagement, which is an essential factor in communicating science with the public.

When it comes to infographics, visual appeal can come in various ways. For example, an infographic may be aesthetically beautiful with a harmonious color palette and an elegant layout. Or, it may look interesting and entertaining by using humor (such as comical drawings), using visual metaphors (like comparing the biological systems of the human body to a subway map), or using unique data visualization (something *other than* bar or pie graphs, as we will see later).

That said, communication remains the primary function of an infographic, and attempts to make it visually pleasing should not impede that function. Consider Figure 7.9, an infographic that illustrates the concept of food hubs, which are businesses that bring local farmers and suppliers together to sell and distribute their products. This infographic features pleasant and appropriately themed earth tone colors. It uses pictorial icons; disarming, non-technical-looking graphs; and unique circular shapes. At a glance, I would say this is a visually attractive design.

Yet, as soon as readers attempt to access the information in this infographic, they will realize that all the texts wrapped around the circular shapes are very difficult, if not impossible, to read. Sure, if someone *really* wants to and doesn't mind spending the time and effort, they will be able to figure the words out, but most readers will not be this motivated. It may seem astonishing that any designer would purposefully design something unreadable, until we realize that this is someone preoccupied with the look of the infographic.

That preoccupation is why efforts to make infographics beautiful can often be counterproductive. It is useful to remember, as mentioned in Chapter 4, that "interest" is not a mere emotional arousal toward things that look appealing. Instead, or in addition, we can interest readers by providing meaningful and compelling information and helping them understand that information. When they do, readers will derive a positive attitude toward the information.

## Putting Numbers in Large Fonts Doesn't Good Infographics Make

Because infographics are frequently used to emphasize numerical data, there is the misconception that if we simply add numbers and highlight them with big fonts, we will be creating something along the line of good infographics.

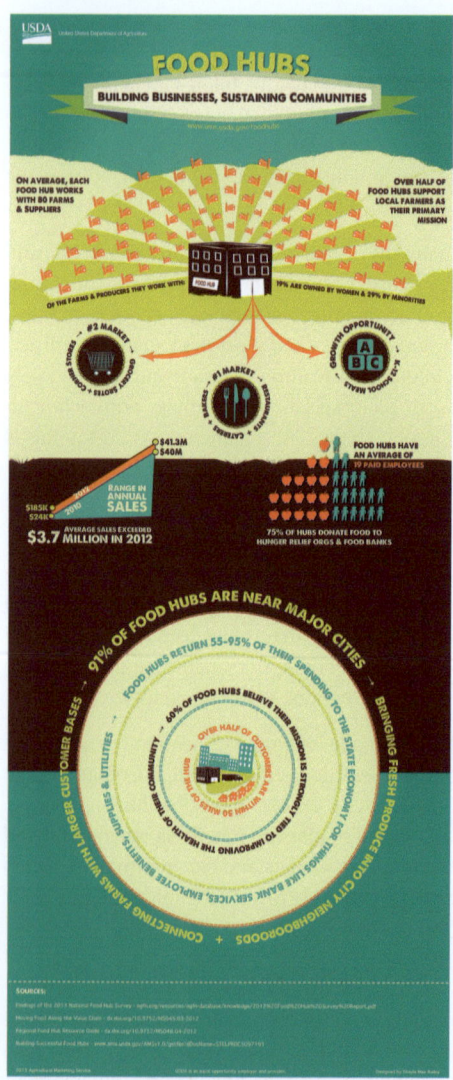

**Figure 7.9**   A seemingly attractive infographic that is difficult to read.

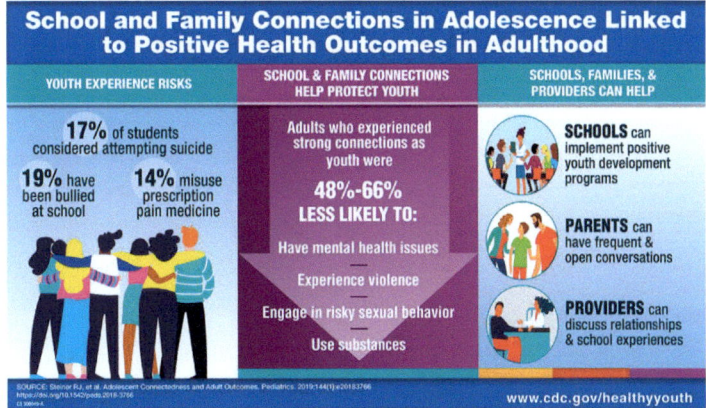

**Figure 7.10** An infographic in which data are not visualized for impact.

Figure 7.10 is an example. It emphasizes the importance of building school and family connections to promote mental wellness and healthy development in youth. Several pieces of numerical data are featured: the percentage of students who have considered suicide, the percentage who were bullied at school, the percentage who misused prescription pain medicine, and the percentage of adults who are less likely to have mental health issues due to school and family connections in their youth.

All of these are relevant data, but none of them are actually visualized or stand out. Although in a somewhat bigger and bolder font, they are merely written out and allowed to blend into other text. Moreover, this treatment of data doesn't help readers remember or appreciate data. For example, 14% of youth misused prescription pain medicine. So? What are we to make of that? Is that somewhat high, high, very high?

To make the data more impactful and memorable, we can use icon arrays (see Chapter 5) to visualize them. Even better, we can put the data into some kind of context or perspective to tell stories. For example, if historical data are available, we can create chronological comparisons to demonstrate how mental wellness and health development in youth

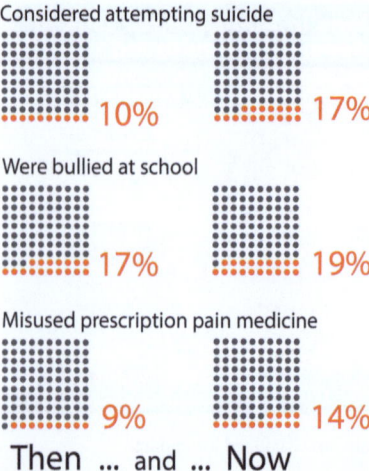

Considered attempting suicide

10%          17%

Were bullied at school

17%          19%

Misused prescription pain medicine

9%          14%

Then ... and ... Now

**Figure 7.11**    An icon array that visualizes and contextualizes data.

have changed over time. If the contemporary situation has worsened, then chronological comparison can help highlight the gravity of the situation and make a more urgent call for action. Figure 7.11 demonstrates such an example (historical data are fictional and used for illustrative purposes). Or, we can compare Figure 7.10 data, which reflect US statistics, with those from other countries. Geographical comparison can enrich readers' understanding of the topic. If the US does better than other countries, readers will realize that the matter is of global concern and appreciate the effort that is currently underway in the US and the need to continue that work. If the US does worse than other countries, readers will appreciate the importance and urgency to take action.

That is all well and good, you say, but what if such comparative data are *not* available? Well, then we go back to some of the earlier lessons: Namely, it is up to the infographic designers to do the (hard) work of creating other meaningful stories about data. If that work isn't done or simply can't be done, then an infographic isn't the right venue for the data.

## Capitalizing Everything Doesn't Good Infographics Make

In document design, capitalization can be used, judiciously, to create visual emphasis. Capitalized letters are bulkier than lowercase letters, so they visually stand out. Unlike lowercase letters, which have distinct shapes (e.g., the letter "a" looks different from the letter "b"), all capitalized letters have the same rectangular shape, which slows down reading and forces readers to pay more attention. This is why we sometimes use capitalized letters for titles and keywords such as "do NOT buy this product!"

Because infographics are only supposed to use small amounts of essential text, some designers assume it is a good idea to capitalize all the text for emphasis, and we end up with infographics like Figure 7.12, which explains the function of fluoride in dental health. In this example, capitalization is used for headings, keywords, and, essentially, all text. Ironically, in the one place where capitalization may be appropriate, the title, it is not used.

The problem is that when everything is capitalized, everything fights for attention, and nothing ends up being emphasized. The extensively capitalized text creates the visual impression of shouting, which is rarely appropriate for an infographic. These texts are also difficult to read (because of the uniform letter shape), making skimming the content difficult for readers, who may be tempted to give up.

Extensive capitalization reduces readability in another way. Capitalized letters take up more physical space, which makes it harder to enlarge the font size. Look at these two words: size and SIZE. The first one is in lowercases and set in a larger font, while the second one is in all caps and set to a smaller font, yet they occupy roughly the same amount of physical space. In other words, if we opt for lowercases, we will have more design space to make fonts larger and easier to read. In Figure 7.12, some of the text is difficult to read not only because it is in all caps but also because it is quite small.

Figure 7.12 may be a pretty extreme example, but an excessive fondness for capitalization *is* prevalent in infographic design. Take, for example,

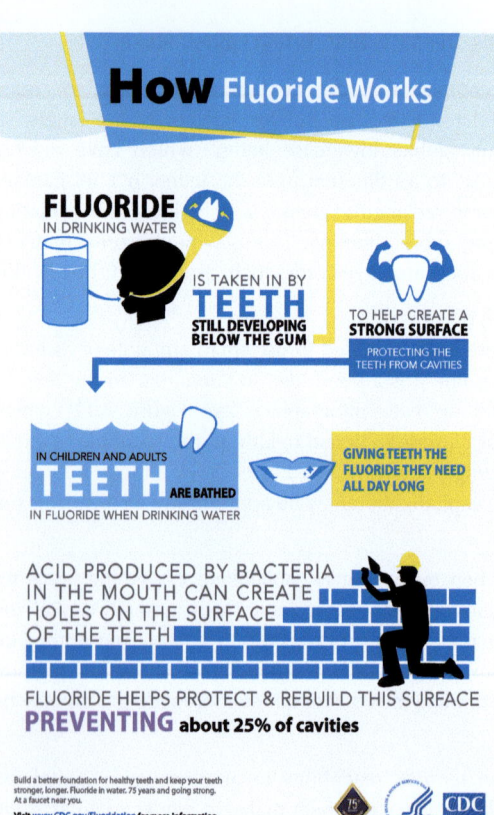

**Figure 7.12**   An infographic full of capitalization.

the food hub infographic we saw in Figure 7.9: Did you notice that all of its text is capitalized too? Or look at Figure 7.8 about the CDC's fight against global tuberculosis: the top two-thirds of it is pretty much all capitalized. If you spend some time browsing the internet for info-graphics, you will see plenty of examples where we can stand to lose some, or a lot, of capitalization.

**Figure 7.13**     Readers prefer pictorial representations of data over standard Cartesian graphs.

## A Sloppy Scale Doesn't Good Infographics Make

As mentioned in Chapter 5, data visualization theorist Edward Tufte resents graphs that use multi-dimensional images (such as dollar bills or oil barrels) to compare one-dimensional data (such as cost or price). He calls these graphs chartjunk. Junk or not, readers in Stones and Gent's interviews liked them and preferred them over standard Cartesian graphs, which they thought were boring. Given the two graphs in Figure 7.13 showing the increase in average male adult weight from the 1960s to the 2020s (created after Stones and Gent; data are hypothetical), readers preferred the one on the right that uses pictorial images to represent data.

It makes sense that readers would prefer the stick figure version. It is more interesting to look at, and it is more intuitive, as it depicts the data it represents (i.e., uses sizes of human figures to represent human weight). The challenge with such graphs, though, is to figure out how to convert one-dimensional data such as weight to two-dimensional pictorial images.

It may be easier to think through this challenge using the good old bar graph. Each bar in a bar graph is essentially a two-dimensional rectangle. It is only by keeping one of the dimensions (the width of the rectangle) consistent that we can then use the other dimension (the height of the rectangle) to compare one-dimensional data. If the widths of the bars differ, then the heights need to be recalculated to represent the same data. Figure 7.14 demonstrates this effect.

In Figure 7.14(a) we have a conventional bar graph where the two bars have the same width. Bar B is three times as tall as bar A, signaling that Group B is three times as large as Group A. Now, if we compare the *sizes* of these two

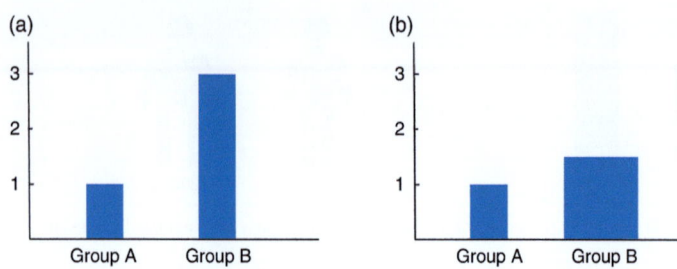

**Figure 7.14** Use two-dimensional images to represent one-dimensional data.

bars, bar B is also three times as large as bar A. The math is simple: bar A is 10 mm wide and 14 mm high, so its size comes to $10 \times 14 = 140 \text{ mm}^2$; bar B is 10 mm wide and 42 mm high, so its size comes to $10 \times 42 = 420 \text{ mm}^2$.

Figure 7.14(b) represents the same data. Bar A has the same width (10 mm), but the width of bar B has doubled to 20 mm. With this setup, we must adjust the height of bar B to 21 mm to ensure that its size remains at 420 mm². That is, $20 \times 21 = 420 \text{ mm}^2$.

The problem is, when we use irregularly shaped pictorial images, this mathematical calculation becomes difficult to apply. In Figure 7.13 I had to apply the math to the "trunk" area of the stick figures and treat that area as a rectangle to create a reasonably approximate pictorial scale.

When designers do not pay close attention to the scales used in pictorial representations of data, we get something like Figure 7.15. This infographic uses interesting visuals, bright colors, and a limited amount of text. It has plenty of white space. It's logically organized and tells an engaging story: Namely, between the 1950s and now, restaurant meal sizes are increasing, and we are gaining weight, and that has to stop. By all the standards we have looked at, this is an excellent infographic – until, that is, we look closely at its scale.

As the opening text in the infographic states, the average restaurant meal size today is more than four times larger than that in the 1950s. The weight labels embedded in the infographic provide more specific comparisons. For

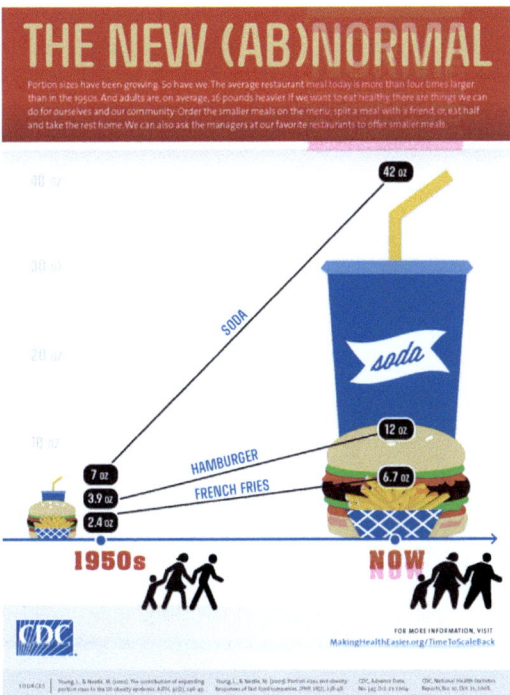

**Figure 7.15**  An infographic with a sloppy scale.

example, a drink grew from 7 oz to 42 oz, so today's drink is 6 times that of the 1950s. However, if we simply eyeball (let alone trying to calculate) the two soda cups, today's cup size is much more than 6 times that of the 1950s' size. The same is true with the burgers, which grew from 3.9 oz to 12 oz, so today's burger is about 3 times that of the 1950s. However, the size difference of the two burgers is far larger than that.

What happened here is that the designer simply relied on the heights of the cups and burgers and fries to denote data, drawing upon the scale on the left side of the infographic. In other words, the pictorial icons were treated as bars with a uniform width. But, clearly, they do not have the same width. When both width and height differ, the size difference – and thus the implied data difference – is exaggerated. It doesn't matter that the scale is there for readers to see. As mentioned in Chapter 5, overall visual impression trumps specific data. When a reader looks at the pictorial images, their mind's eye will automatically compare the sizes of the images. And this is why Tufte calls these graphs lying graphs.

If Tufte had his way, he would probably ban all such data visualization from infographics, and many would probably agree. Personally, I concur that using *three*-dimensional pictorial images to denote data is to be avoided: If converting one-dimensional data to two-dimensional images is already complicated, imagine how messy it would be to work with three-dimensional images. On the other hand, simple two-dimensional images, like those used in Figures 7.13 and 7.15, can be fun and intuitive for readers – as long as designers are aware of the issue of the pictorial scale. Granted, the result will not be as exact as standard Cartesian graphs, because precise scaling is nearly impossible with irregular shapes. But ultimate precision is often not a goal in popular science communication. As with pictographs (see Chapter 5), to be able to remember simplified pictures is better than forgetting precise numbers.

## Conclusion

Infographics, with their multiple and juxtaposed visuals and texts, are superb at telling stories and making calls for actions. Used well, they can be an effective tool to communicate science to the public. At the same time, because infographics contain multiple visuals and texts, they also face increased complications. Misconceptions concerning standalone visuals that were discussed in previous chapters often apply to infographics, such as an overemphasis on visual appeal or the temptation to use superficially relevant visuals. In addition, infographics face unique challenges because they have to juggle multiple visual and textual elements, which demand careful consideration of placement, organization, typography, and more.

Designers of infographics need to be extra conscious – and conscientious – of their design choices. Creating infographics is more than dumping some numbers on a template. It takes a tremendous amount of work and time to ensure that the visuals, the text, the organization, and the embedded stories actually speak to a target audience and are deemed appealing and easily understandable by the audience.

# Conclusion

Writing this book has been a pleasure. The scholarly significance of it aside, I just enjoy visuals. I enjoy looking at them, thinking about them, and talking about them. And that's not limited to my research interest in science visuals either. I like all of them.

I delight, for example, in IKEA's visual-only assembly instructions (I'm fully aware that some people don't). For me, I look at them, I look back at the mess lying on the floor, I see the panels I need, the screws I need, I put them together step by step, just as the visuals show. Everything feels easy, efficient, and intuitive. Just as much, I hate it when I stumble upon a bad set of visual instructions. Examples of those are plentiful. The most recent memory pertains to putting up the bagger kit on our Greenworks mower. Great, great mower really (and great bagger, too), but the instructions? Not so much. The photographs are grainy, dark, and small, so I struggled to see what's on them and to identify the parts of importance. The angle of the photographs wasn't quite right, so I *really* struggled to see what I'm supposed to do with the parts. It took a lot of trial and error, a lot of straining, and, I'm not proud to say, a fair amount of swearing.

So, here's my point: Visuals are integral to our lives, our understanding of the world, and our interactions with that world. For disciplines that endeavor to study the world and lives within that world, visuals are a paramount tool.

Every life scientist and communicator who has occasion to create visuals for nonspecialist readers, I hope, can learn to appreciate visuals, to appreciate thinking and talking about them. As I tried to show in this book, there truly is a lot to these visuals, if we want to do right by our readers. These visuals cannot be taken for granted as mere decorations of text, signifying nothing. Conversely, they cannot be smokescreens that hide complicated knowledge that a writer finds difficult (or inconvenient, or whatever) to actually explain. These visuals need to be interesting and relatable and perhaps awe-inspiring when the occasion calls for it, but they also must be genuine, relevant, and relay the right kind and amount of information.

This, of course, is easier said than done – almost everything worth doing is easier said than done – but, it certainly can be done. I believe that what it takes, as with most things, is dedication and practice. We need to dedicate ourselves to studying the unique needs and wants of nonspecialist readers when it comes to science visuals. We also need to be diligent in our visual creation process. Next time you attempt to create a visual, think about what you wish to convey and what visual features are consequently required before you start drawing, photographing, or sourcing. And when you do get something on paper, be very critical about what it shows, what it doesn't show, and how it shows – be as critical as you are when conducting science or writing about science. Consider if anything needs to be changed to serve your purpose, your audience, better. Be brave enough to toss the whole thing out and start afresh, if that's what's needed.

If you are (at least sometimes) a nonspecialist reader on the other side of science communication, I hope this book has proven useful to you as well. The next time you stare at a visual put out by, say, the CDC, and you just don't know what you are looking at, I hope you realize that it is *not* your fault, your lack of knowledge, or your lack of "science smart." Rather, that visual isn't right for you or hasn't been made right for you. I hope that you feel empowered to ask difficult questions instead of taking a pretty micrograph for granted or brushing a complicated graph aside.

We live in a vibrantly visual world with dazzling life forms, it is only appropriate that our studies of them are purposefully visual.

# Summary of Common Misunderstandings

**A staple of everyday life, photographs make accessible life science visual evidence.** Until the nineteenth century, life scientists were primarily concerned with fact-gathering of the natural world as opposed to theory building. Because of this, photographs were a superior type of visual evidence: They recorded the physical appearance of nature in ways that far surpassed the written word. Moreover, photographs reflected the scientific method behind the study: that of gathering and cataloging. As a result, what you see in the photographs is pretty much what you get, and what you get is the precise value of the scientific work. This, however, is no longer the case in contemporary life science research. The goal of contemporary research shifted from documenting what life looks like to unveiling the why and how of life forms. Research methods also shifted from gathering and cataloging subjects to experimenting with subjects. As a result, the value of photographs as scientific records and evidence diminished. Their significance as research evidence also became far less transparent to a nonspecialist audience.

**Photographs and micrographs are made by machines and record objective reality.** People who subscribe to the concept of "mechanical objectivity" believe that the best scientific visuals are the ones that are free of human manipulation and allow nature to speak for itself. Moreover, they believe that to create mechanically objective visuals we need to rely on machines such as cameras. As machines, cameras have no personal agenda or interest. The resultant photographs (or micrographs, when cameras are outfitted on microscopes) are therefore objective, superior to human-made drawings that are the result of their makers' bias and imagination. This line of reasoning forgets that machines are set and operated by humans. In the case of cameras, anything from lens selection to exposure time can affect the photographic outcome. There is also the purposeful selection of what scenes to capture, what scenes to ignore, and post-editing enabled by software. Objectivity, in the mechanical sense of the word, doesn't exist, and any photograph or micrograph is a result of human intervention. Indeed, in popular science communication they are frequently endowed with emotions and values, sometimes, perhaps, to the extreme.

**Micrographs reveal the microscopic world invisible to the public, and therein lies their value.** Throughout the history of microscopes, we pursued inventions that allowed us to see ever more minutely and clearly. We assume that when we break nature down to its smallest parts and behold those parts, we *will* understand. Precisely because the microscopic world is invisible in everyday life, we assume that when the public behold that world, they will automatically appreciate the value and significance of life science research. Such a reductive focus can pose a challenge in popular science communication. Because microorganisms have no counterpart in the everyday visual world, there is little external reference that nonspecialist viewers can use to comprehend a micrograph. Therefore, a "look at X" micrograph can easily become a "look for X" exercise. More importantly, without contextual cues and intelligent stories, a static, ultra-zoomed-in micrograph holds very little visual significance for the public.

**Illustrations look through the appearance of nature to reveal life as it really is.** Scientific illustrations are often not concerned with the rich details of a subject's appearance. In fact, they deem such details distracting. Instead, illustrations aim to foreground a subject's most essential visual elements: the how, the why, the inner workings, the causal structures. The clean lines, sharp edges, and concrete icons all contribute to this mission. It is, then, easy to forget that illustrations are the results of active interpretation, that they are the products of our beliefs, even our imagination. By committing ink to paper, illustrations eradicate any doubt, confusion, and debate there may be. They downplay the fact that what counts as "essential" differs from one visual creator to another. They also downplay the fact that contemporary illustrations are full of pseudo-details. The vivid textures, rich colors, and multi-dimensional shapes are but a version of reality, not *the* reality. At their worst, these details illustrate very little about the workings and structures of life.

**Graphs present "hard" numbers, and numbers don't lie.** Doctored numbers are lies, that much we can all agree. But even unaltered numbers can tell different stories, depending on how we graph them. Truncating the $y$-axis so that dependent variables start not at zero but at an arbitrary number exaggerates the differences between data. Alternatively, setting a high maximum bound on the $y$-axis flattens differences and hides fluctuations in data. Using a logarithmic scale versus an arithmetic scale dramatically changes the appearance of data patterns. Different variations of the bar graph emphasize different kinds of group and subgroup comparisons, changing our perspectives about data. All of these practices and more may not be labeled "lies," but they certainly highlight the fact that graphs are very much a rhetorical device.

**Pictographs are chartjunk and unfit for science communication.** Data visualization theorist Edward Tufte is famous for calling graphs that use decorative details chartjunk. In particular, Tufte criticizes graphs that use two- or three-dimensional pictorial images to represent one-dimensional data, such as using dollar bills of different sizes to chart the rate of inflation. These critiques deter the use of pictographs in popular science communication. The assumption that pictographs are fit only for children and inappropriate for serious occasions contributes to their unpopularity. But, in reality, pictographs work fundamentally differently from Tufte's examples of lying graphs. Pictographs employ icons such as human figures, everyday objects, and simple geometric shapes. The number, not the area or volume, of these icons encodes data. Pictographs are a missed opportunity for popular science communication because they offer multiple benefits: They add visual interest and attract viewer attention; they make data more intuitive, concrete, and easier to grasp; and, designed well, they can tell rich stories about data.

**Interactive visualization technologies equate to effective popular science communication.** Contemporary life sciences are big data sciences. When it comes to analyzing and presenting big data, interactive online visuals – maps, graphs, three-dimensional models – have inherent advantages. They are dynamic and easily updated. They support user interaction and allow users to filter data and create displays that make sense to *them*. That said, just because something is online and interactive or uses the latest technologies does not necessarily mean that it is engaging and meaningful. Despite the power and versatility of interactive visuals, they can't gracefully handle any and all data thrown at them. Too many buttons, menu options, and layers of data complicate the user interface and paralyze users. Even more importantly, we need thorough research on what public users actually desire from interactive visualization technologies.

**Visual appeals equate to effective popular science communication.** In contemporary life, we have many (too many, perhaps) entertainment options right at our fingertips. From HBO shows to video games to blockbuster movies, there is plenty out there to vie for public attention, to, in other words, compete with popular science communication. So, the pressure is on for scientists and science communicators to make science appear interesting and appealing. Efforts toward this goal include, among other things, stunning micrographs, elaborate illustrations, and interactive games. There is no doubt that surface-level attraction is imperative. After all, if people are not looking, there can be no communication. But, at the same time, surface-level attraction doesn't equate to effective communication or engagement. In fact, at their worst, surface-level attraction impedes effective communication and engagement. It can lead people to believe that science is simple, familiar, and *not* worth asking specific questions, that playing a certain computer game, as one *WIRED* magazine reporter writes, makes them a genetic scientist. It most certainly does not.

**A mixture of visuals, numbers, and text results in an infographic.** Infographics are, indeed, a unique multimodal genre that combines visuals, numbers, and text. But the reverse isn't true. That is, simply mixing up visuals, numbers, and text on a page doesn't necessarily result in an infographic, at least not a good one. Many design guidelines go into choosing, presenting, and organizing the right kinds of visuals, numbers, and text. For example, visuals need to convey critical information, not merely be of superficial relevance, and numbers need to be visualized and contextualized, not merely written out in large fonts as isolated pieces of information. Care needs to be taken to create a clean design that allows important messages to stand out. Moreover, visuals, numbers, and text need to be well organized and work together to tell meaningful stories.

# Figure Credits

Figure 1.1    Hooke, R. (1665). *Micrographia: Or some physiological descriptions of minute bodies made by magnifying glasses. With observations and inquiries thereupon.* London: Royal Society. Plate XXXIV. Retrieved October 29, 2023, from https://digital.science history.org/works/9g54xj51s/viewer/bk128c17q

Figure 1.2    Vesalius, A. (1543). *De humani corporis fabrica libri septem* (p. 221). Source: Wellcome Collection. Basileae: Per Joannem Oporinum. Retrieved December 3, 2023, from https://wellco mecollection.org/works/mv74d54w/items?canvas=243

Figure 1.3    Pereyra, S., Sosa, C., Bertoni, B., & Sapiro R. (2019). Transcriptomic analysis of fetal membranes reveals pathways involved in preterm birth. *BMC Med Genomics, 12*(53). https:// doi.org/10.1186/s12920-019-0498-3. Open Access article, CC BY 4.0. *Copyright © 2019, The Author(s).*

Figure 2.1    Atkins, A. (1849–1850). *Photographs of British algae: Cyanotype impressions.* Retrieved November 4, 2023, from https://digitalcollections.nypl.org/items/510d47d9-4aef-a3d9-e040-e00a18064a99

Figure 2.2    Duchenne de Boulogne, G. B. (1862). *Mécanisme de la physionomie humaine.* Retrieved October 22, 2023, from https:// publicdomainreview.org/collection/the-mechanism-of-human-physiognomy

Figure 2.3    United States Surgeon General. (1865). *Photographs of surgical cases and specimens.* Volume 1. Retrieved November 3, 2023, from https://archive.org/details/photographsofsur01unit/page/n 93/mode/2up

Figure 2.4    Franklin, R. & Gosling, R. G. (1953). Molecular configuration in sodium thymonucleate. *Nature, 171*(4356), 740–741. Reprinted by permission from Springer Nature.

Figure 2.5    Dolinoy, D. C., Weidman, J. R., Waterland, R. A., & Jirtle, R. L. (2006). Maternal genistein alters coat color and protects $A^{vy}$ mouse offspring from obesity by modifying the fetal epigenome. *Environmental Health Perspectives, 114*(4), 567–572.

Figure 2.6    Southeast Fisheries Science Center Pascagoula Laboratory (2019). Collection of Brandi Noble, NOAA/NMFS/SEFSC. Retrieved December 3, 2023, from https://photolib.noaa.gov/Collections/Fisheries/Life-In-The-Sea/Southeast-Fisheries-Science-Center/emodule/1062/eitem/47294

Figure 2.7    National Oceanic and Atmospheric Administration Monterey Bay National Marine Sanctuary (n.d.). Retrieved December 3, 2023, from https://nmssanctuaries.blob.core.windows.net/sanctuaries-prod/media/eib/2265.jpg

Figure 2.8    Frank, M. (2013). Photograph of a mare's (female horse) uterus with the fetus removed. Attribution-NonCommercial 4.0 International (CC BY-NC 4.0). Source: Wellcome Collection. Retrieved December 3, 2023, from https://wellcomecollection.org/works/ux6mpb4v

Figure 2.10    United States Surgeon General. (1865). *Photographs of surgical cases and specimens*. Volume 5. Retrieved November 3, 2023, from https://archive.org/details/photographsofsur05unit/page/n21/mode/2up

Figure 2.11    Doyen, E.-L. (1911). *Atlas d'anatomie topographique*. Paper 7. Plate 31. Retrieved December 3, 2023, from https://collections.nlm.nih.gov/catalog/nlm:nlmuid-101595354-img

Figure 3.1    Hooke, R. (1665). *Micrographia: Or some physiological descriptions of minute bodies made by magnifying glasses. With observations and inquiries thereupon*. London: Royal Society. Plate XXXV. Courtesy of Science History Institute. Retrieved December 8, 2023, from https://digital.sciencehistory.org/works/9g54xj51s/viewer/jq085m095

Figure 3.11   Häggström, M. (2019). Micrograph of colorectal carcinoma with dirty necrosis. Retrieved November 8, 2023, from https://commons.wikimedia.org/wiki/File:Micrograph_of_colorectal_carcinoma_with_dirty_necrosis.jpg. CC0 1.0.

Figure 3.12   The National Institute of Allergy and Infectious Diseases (2020). Novel Coronavirus SARS-CoV-2. Retrieved November 8, 2023, from www.flickr.com/photos/niaid/49645402917/

Figure 3.13   National Institutes of Health (2020). Novel coronavirus structure reveals targets for vaccines and treatments. Retrieved November 8, 2023, from www.nih.gov/news-events/nih-research-matters/novel-coronavirus-structure-reveals-targets-vaccines-treatments

Figure 3.14   National Institute of Allergy and Infectious Diseases (2014). Using genomics to follow the path of Ebola. Retrieved November 8, 2023, from https://directorsblog.nih.gov/2014/09/02/using-genomics-to-follow-the-path-of-ebola

Figure 3.15   Troemel, E. R., Félix, M.-A., Whiteman, N. K., Barrière, A., & Ausubel F. M. (2008). Microsporidia are natural intracellular parasites of the nematode *Caenorhabditis elegans*. *PLoS Biology*, 6(12): e309. https://doi.org/10.1371/journal.pbio.0060309. CC BY 4.0.

Figure 3.16   Head louse clinging to strand of human hair, SEM. (n.d.) Kevin Mackenzie, University of Aberdeen. Attribution 4.0 International (CC BY 4.0). Source: Wellcome Collection. Retrieved November 8, 2023, from https://wellcomecollection.org/works/abydbmms/items

Figure 4.1   da Vinci, L. (c. 1511). Studies of the fetus in the womb. Retrieved November 13, 2023, from https://en.wikipedia.org/wiki/Studies_of_the_Fetus_in_the_Womb#/media/File:Leonardo_da_Vinci_-_Studies_of_the_foetus_in_the_womb.jpg

Figure 4.2   Sun, S. M. (n.d.). San ren ming tang tu. Retrieved November 13, 2023, from www.baike.com/wikiid/8903893017839555586?view_id=32yxgkosj3bi80. Reproduction.

Figure 4.3   Merian, M. S. (1679). *Der Raupen wunderbare Verwandelung, und sonderbare Blumen-nahrung*. Bd. 1. Nürnberg 1679. Abbildung 1 (Maulbeerbaum samt der Frucht). Kolorierter

Stich. Retrieved November 13, 2023, from https://commons
.wikimedia.org/wiki/File:Merian_-_Der_Raupen_wunderbare_
Verwandelung_und_sonderbare_Blumennahrung_-_Abb_1.jpg

Figure 4.4    Belomaad (2021). Basic structures of the brain highlighted.
Retrieved November 13, 2023, from https://commons.wikime
dia.org/wiki/File:Basic_structures_of_the_brain_highlighted.png.
CC BY-SA 4.0.

Figure 4.5    Anatomical structure of biological animal cell with organelles.
(n.d.). ©eranicle/123RF.COM. Retrieved November 13, 2023,
from www.123rf.com/photo_55829381_anatomical-structure-
of-biological-animal-cell-with-organelles.html

Figure 4.6    (a) Openstax (1999–2024). *Biology*: *The Central Nervous System*.
Retrieved November 13, 2023, from https://openstax.org/books/
biology/pages/35-3-the-central-nervous-system. CC BY 4.0.

Figure 4.7    Melvin, C. (1975). Chemical evolution: Life is a logical conse-
quence of known chemical principles operating on the atomic
composition of the universe. *American Scientist*, *63*(2), 169–177.

Figure 4.8    Getty Images. (n.d.). Protein synthesis drawing. Retrieved
December 2, 2023, from https://media.gettyimages.com/id/14
25532489/fr/vectoriel/protein-synthesis-drawing.jpg?s=1024
x1024&w=gi&k=20&c=6e8WJQcjXR5vrsjTcqvUzYCZ5uTy9P
R1xoF6kqwUlqY=

Figure 4.9    Alissa Eckert, MSMI, Dan Higgins, MAMS. Centers for Disease
Control and Prevention. (2020). Retrieved December 13, 2023,
from https://phil.cdc.gov/details.aspx?pid=23312

Figure 4.10   Research Collaboratory for Structural Bioinformatics Protein
Data Bank (2020). Structure of the SARS-CoV-2 spike glycopro-
tein (closed state). Retrieved March 19, 2024, from www.rcsb
.org/structure/6vxx

Figure 4.11   National Institutes of Health (2009). New NIH research plan on
Fragile X Syndrome and associated disorders. Retrieved
November 13, 2023, from www.nichd.nih.gov/newsroom/
resources/spotlight/071609-Fragile-X

Figure 4.12   Courtesy: National Human Genome Research Institute (2023).
Chromosome. Retrieved November 13, 2023, from www
.genome.gov/genetics-glossary/Chromosome

Figure 4.13    Dhama, K., Khan, S., Tiwari, R., et al. (2020). Coronavirus disease 2019: COVID-19. *Clinical Microbiology Reviews*, *33* (4), e00028-20. Reproduced with permission from American Society for Microbiology, permission conveyed through Copyright Clearance Center, Inc.

Figure 4.14    Genetic manipulation and DNA modification concept. (n.d.). ©vchalup/123RF.COM. Retrieved November 13, 2023, from www.123rf.com/photo_85839783_genetic-manipulation-and-dna-modification-concept.html

Figure 4.15    Abstract model of man of DNA molecule. (n.d.). ©lonely11/123RF.COM. Retrieved November 13, 2023, from www.123rf.com/photo_13715924_abstract-model-of-man-of-dna-molecule.html

Figure 4.16    Ball, P. (2016). CRISPR: Implications for materials science. *MRS Bulletin*, *41*, 832–835. Reprinted by permission from Springer Nature.

Figure 5.1    Data source: United Nations Department of Economic and Social Affairs; Gapminder. Retrieved November 13, 2023, from www.statista.com/statistics/1042370/united-states-all-time-infant-mortality-rate

Figure 5.2    Data source: United Nations Department of Economic and Social Affairs; Gapminder. Retrieved November 13, 2023, from www.statista.com/statistics/1042370/united-states-all-time-infant-mortality-rate

Figure 5.3    Centers for Disease Control and Prevention (2020). Daily number of reported COVID-19 cases – United States, February 12–March 28, 2020. Retrieved November 13, 2023, from www.cdc.gov/mmwr/volumes/69/wr/mm6913e2.htm

Figure 5.4    Centers for Disease Control and Prevention (2011). Retrieved November 13, 2023, from www.cdc.gov/nchs/images/databriefs/101-150/db115_fig4.png

Figure 5.5    Centers for Disease Control and Prevention (2014). The geography of diabetes by census tract in a large sample of insured adults in King County, Washington, 2005–2006. Retrieved November 13, 2023, from www.cdc.gov/pcd/issues/2014/14_0135.htm#table2_down

Figure 5.7    Romano, A., Sotis, C., Dominioni, G., & Guidi, S. (2020). The scale of COVID-19 graphs affects understanding, attitudes, and policy preferences. *Health Economics*, *29*(11), 1482–1494. © 2020 The Authors. *Health Economics* published by John Wiley & Sons Ltd. CC BY 4.0.

Figure 5.8    Pew Research Center (2014). American Trends Panel Wave 6. Retrieved November 13, 2023, from www.pewresearch.org/politics/dataset/american-trends-panel-wave-6

Figure 5.10   Created based on the format used in Mitropoulos, A., Brownstein, J., & Bhatt, J. (2021). COVID-19 vaccinations finally starting to stem pandemic's tide in US: Analysis. Retrieved November 13, 2023, from https://abcnews.go.com/Health/covid-19-vaccinations-finally-starting-stem-pandemics-tide/story?id=77514759. Data are illustrative, not factual.

Figure 5.11   Data source: United Nations Department of Economic and Social Affairs; Gapminder. Retrieved November 13, 2023, from www.statista.com/statistics/1042370/united-states-all-time-infant-mortality-rate

Figure 5.12   Centers for Disease Control and Prevention (2012). By race/ethnicity: African Americans are least likely to be in ongoing care or to have their virus under control. Retrieved November 13, 2023, from www.cdc.gov/nchhstp/newsroom/images/2012/Stages-of-CareFig2.jpg

Figure 5.13   Centers for Disease Control and Prevention (2009). Increase in fatal poisonings involving opioid analgesics in the United States, 1999–2006. Source: CDC/NCHS, National Vital Statistics System. Retrieved November 13, 2023, from www.cdc.gov/nchs/products/databriefs/db22.htm

Figure 5.14   Centers for Disease Control and Prevention (2021). County-level COVID-19 vaccination coverage and social vulnerability – United States, December 14, 2020–March 1, 2021. Retrieved November 13, 2023, from www.cdc.gov/mmwr/volumes/70/wr/mm7012e1.htm

Figure 5.15   McCarthy, N. (2019). Milk's massive American decline. Retrieved November 13, 2023, from www.statista.com/chart/2387/american-milk-consumption-has-plummeted. CC BY-ND.

Figure 5.16    Centers for Disease Control and Prevention (2019). NCHS mortality surveillance data. Retrieved November 13, 2023, from https://archive.cdc.gov/#/details?url=https://www.cdc.gov/coronavirus/2019-ncov/covid-data/covidview/10302020/nchs-mortality-report.html

Figure 5.17    Remington, P. L., Catlin, B. B., & Kindig, D. A. (2013). Monitoring progress in population health: Trends in premature death rates. *Preventing Chronic Disease, 10.* Retrieved November 13, 2023, from www.cdc.gov/pcd/issues/2013/13_0210.htm

Figure 5.18    Based on data from Statista: Leading cotton producing countries worldwide in 2020/2021. Retrieved January 2, 2022, from www.statista.com/statistics/263055/cotton-production-worldwide-by-top-countries

Figure 5.19    *Gesellschaft und Wirtschaft: Bildstatistisches Elementarwerk.* (1930). Leipzig: Verlag des Bibliographischen Instituts AG. P. 74. Wienbibliothek im Rathaus. Retrieved November 13, 2023, from www.digital.wienbibliothek.at/urn/urn:nbn:at:AT-WBR-125389

Figure 5.20    Graph developed using Icon Array (www.iconarray.com).

Figure 6.1    Centers for Disease Control and Prevention (2024). United States COVID-19 hospitalizations, deaths, emergency department (ED) visits, and test positivity by geographic area. Retrieved November 15, 2023, from https://covid.cdc.gov/covid-data-tracker/#maps_percent-covid-ed

Figure 6.2    Based on data from Johns Hopkins University, Coronavirus Resource Centre (2023). COVID-19 United States cases by county. Retrieved March 4, 2022, from https://coronavirus.jhu.edu/us-map

Figure 6.3    Based on data from Johns Hopkins University, Coronavirus Resource Centre (2023). COVID-19 United States cases by county. Retrieved March 4, 2022, from https://coronavirus.jhu.edu/us-map

Figure 6.4    Centers for Disease Control and Prevention (2023). Trends in United States COVID-19 hospitalizations, deaths, emergency department (ED) visits, and test positivity by geographic area.

Retrieved November 15, 2023, from https://covid.cdc.gov/covid-data-tracker/#trends_totaldeaths_select_00

Figure 6.5    Centers for Disease Control and Prevention (2023). Trends in United States COVID-19 hospitalizations, deaths, emergency department (ED) visits, and test positivity by geographic area. Retrieved November 15, 2023, from https://covid.cdc.gov/covid-data-tracker/#trends_totaldeaths_weeklyhospitaladmissions100k_00

Figure 6.6    Based on data from Johns Hopkins University, Coronavirus Resource Centre (2023). Cumulative cases by days since 50th confirmed case. Retrieved March 4, 2022, from https://coronavirus.jhu.edu/data/cumulative-cases

Figure 6.7    Based on data from Johns Hopkins University, Coronavirus Resource Centre (2022). Impact of opening and closing decisions by state. Retrieved November 15, 2023, from https://coronavirus.jhu.edu/data/state-timeline/new-confirmed-cases/illinois/46

Figure 6.8    91-DIVOC. (n.d.). An interactive visualization of the exponential spread of COVID-19. Retrieved March 4, 2022, from https://91-divoc.com/pages/covid-visualization/. CC BY 4.0.

Figure 6.9    91-DIVOC. (n.d.). An interactive visualization of the exponential spread of COVID-19. Retrieved March 4, 2022, from https://91-divoc.com/pages/covid-visualization/. CC BY 4.0.

Figure 6.10   Created by author using Phylo. McGill (2023). Retrieved March 4, 2022, from https://phylo.cs.mcgill.ca

Figure 6.11   Created by author using Foldit. (n.d.). Retrieved March 4, 2022, from https://fold.it

Figure 6.12   Cooper, S., Khatib, F., Treuille, A., et al. (2010). Predicting protein structures with a multiplayer online game. *Nature*, *466*(7303), 756–760. Supplementary information. Retrieved November 21, 2023, from https://static-content.springer.com/esm/art%3A10.1038%2Fnature09304/MediaObjects/41586_2010_BFnature09304_MOESM302_ESM.pdf. Figure S2. Reprinted by permission from Springer Nature.

Figure 6.13   Herráez, A. (n.d.). DNA structure tutorial. Retrieved May 5, 2022, from https://biomodel.uah.es/en/model4/dna/dnapairs.htm. CC BY-NC-SA.

Figure 6.14   National Library of Medicine (n.d.). Retrieved November 16, 2023, from https://data.lhncbc.nlm.nih.gov/public/Visible-Human/Female-Images/PNG_format/legs/avf2206a.png

Figure 7.1   United States Department of Agriculture (2020). Brown rice 5 ways. Retrieved November 16, 2023, from https://myplate-pro d.azureedge.us/sites/default/files/2020-12/Brown%20Rice%2 05%20Ways.pdf

Figure 7.2   Centers for Disease Control and Prevention (2019). Antibiotic resistance: The global threat. Retrieved November 16, 2023, from www.cdc.gov/globalhealth/infographics/antibiotic-resist ance/antibiotic_resistance_global_threat.htm. An excerpt.

Figure 7.3   United States Department of Health & Human Services (2019). Your map to a food-safe holiday. Retrieved November 16, 2023, from www.foodsafety.gov/sites/default/files/2019-05/ winter-holiday-food-safety-infographic.jpg

Figure 7.4   Centers for Disease Control and Prevention (n.d.). Diabetes in the United States. Retrieved November 16, 2023, from https:// khni.kerry.com/news/articles/a-snapshot-of-diabetes-in-the-un ited-states/. An excerpt.

Figure 7.5   United States Department of Agriculture (2020). Tomato. Retrieved November 16, 2023, from https://myplate-prod.azur eedge.us/sites/default/files/2020-12/Tomato%20Fact%20Card %20%282017%29.pdf

Figure 7.6   Centers for Disease Control and Prevention (2023). What is contact tracing? Retrieved November 16, 2023, from www .cdcmuseum.org/exhibits/show/ebola/item/99

Figure 7.7   United States Department of Agriculture (2023). Make every bite count with the *Dietary Guidelines for Americans*. Retrieved November 16, 2023, from www.dietaryguidelines.gov/sites/ default/files/2023-08/DGA-2020-2025-Infographic-MakeEver yBiteCount.pdf

Figure 7.8   Centers for Disease Control and Prevention (2017). CDC's fight against global tuberculosis. Retrieved November 16, 2023, from https://archive.cdc.gov/#/details?url=www.cdc.gov/ globalhealth/infographics/tb/global_tb_2016.htm

Figure 7.9    United States Department of Agriculture (2013). Food hubs: Building businesses, sustaining communities. Retrieved December 4, 2023, from www.usda.gov/media/blog/2013/11/20/food-hubs-building-businesses-and-sustaining-communities

Figure 7.10   Centers for Disease Control and Prevention (n.d.). School and family connections in adolescence linked to positive health outcomes in adulthood. Retrieved November 16, 2023, from www.cdc.gov/healthyyouth/protective/images/youth_connectedness_visual_abstract-large.jpg?_=62834

Figure 7.12   Centers for Disease Control and Prevention (2021). How fluoride works. Retrieved November 16, 2023, from www.cdc.gov/fluoridation/resources/how-fluoridation-works.html

Figure 7.13   Based on data from Stones, C. & Gent, M. (2015). *The 7 G.R.A.P.H.I.C. Principles of Public Health Infographic Design*, p. 33. Retrieved November 16, 2023, from https://visualisinghealth.files.wordpress.com/2014/12/guidelines.pdf. Data are hypothetical.

Figure 7.15   Centers for Disease Control and Prevention (2020). The new (ab)normal. Retrieved November 16, 2023, from www.cdc.gov/nccdphp/dnpao/multimedia/infographics/newabnormal.html

# References

## Preface

It is possible to differentiate "scientific communication" from "science communication." Interested readers can refer to Yu, H. & Northcut, K. (2017). *Scientific Communication: Practices, Theories, and Pedagogies*. New York: Routledge.

For readers interested in the use of visuals in professional and classroom science communication, Luc Pauwels' collection is a good starting point: Pauwels, L. (2006). *Visual Cultures of Science: Rethinking Representational Practices in Knowledge Building and Science Communication*. Lebanon, NH: Dartmouth College Press.

## Chapter 1

Galison, P. (1998). Judgment against objectivity. In C. A. Jones, P. Galison, & A. Slaton (Eds.), *Picturing Science, Producing Art* (pp. 327–359). New York: Routledge.

Holliman, R., Whitelegg, E., Scanlon, E., Smidt, S., & Thomas, J. (2009). Final reflections … . In R. Holliman, E. Whitelegg, E. Scanlon, S. Smidt, & J. Thomas (Eds.), *Investigating Science Communication in the Information Age* (pp. 274–278). Oxford: Oxford University Press.

National Science Board (2018). *Science and Engineering Indicators 2018*. Alexandria, VA: National Science Foundation.

Trumbo, J. (2000). Seeing science: Research opportunities in the visual communication of science. *Science Communication, 21*(4), 379–391.

Wynne, B. (2004). Misunderstood misunderstandings: Social identities and public uptake of science. In A. Irwin and B. Wynne (Eds.), *Misunderstanding Science? The Public Reconstruction of Science and Technology* (pp. 19–46). Cambridge: Cambridge University Press.

## Chapter 2

Bloomfield, B. P., & Doolin, B. (2012). Symbolic communication in public protest over genetic modification: Visual rhetoric, symbolic excess, and social mores. *Science Communication*, *35*(4), 502–527.

Brownlee, C. (2006). Nurture takes the spotlight. *Science News*, *169*(25), 392–393, 396.

Gross, A. G., Harmon, J. E., & Reidy, M. (2002). *Communicating Science: The Scientific Article from the 17th Century to the Present*. New York: Oxford University Press.

Jabed, A., Wagner, S., McCracken, J., Wells, D. N., & Laible, G. (2012). Targeted microRNA expression in dairy cattle directs production of β-lactoglobulin-free, high-casein milk. *Proceedings of the National Academy of Sciences*, *109*(42), 16811–16816.

Kress, G. & van Leeuwen, T. (2006). *Reading Images: The Grammar of Visual Design* (2nd ed.). London: Routledge.

United States Surgeon General (1865). Photographs of surgical cases and specimens. Retrieved November 3, 2023, from https://archive.org/details/photographsofsur01unit

van Gieson, R. E. (1860). The application of photography to medical science, including a direct process to photograph the microscopic field. *New York Journal of Medicine*, *8*(1), 1–17. I'm indebted to Michael Sappol for this source: Sappol, M. (2017). Anatomy's photography: Objectivity, showmanship and the reinvention of the anatomical image 1860–1950. Retrieved April 3, 2023, from https://remedianetwork.net/2017/01/23/anatomys-photography-objectivity-showmanship-and-the-reinvention-of-the-anatomical-image-1860-1950/

The rebranded Wellcome Photography Prize focuses on health and medicine, and winning works have been photographs of humans that have a socially charged visual impact rather than pure visual splendor.

## Chapter 3

Chow, D. (2021). Why scientists are talking about viral load and the delta variant. Retrieved November 22, 2023, from www.nbcnews.com/science/science-ne ws/delta-variant-viral-load-scientists-are-watching-covid-pandemic-rcna1604

Daston, L. & Galison, P. (2007). *Objectivity*. New York: Zone Books.

de Jesús, E. G. (2020). SARS and the new coronavirus target the same cellular lock to infect cells. *Science News*. Retrieved November 22, 2023, from www.scien cenews.org/article/sars-new-coronavirus-target-same-cellular-lock-infect-cells

Hopwood, N. (2015). *Haeckel's Embryos: Images, Evolution, and Fraud*. Chicago, IL: University of Chicago Press.

Lane, M. A. (1908). Weismannism. *Scientific American*, *1698* (Suppl.), 45–46.

National Institutes of Health (2020). Novel coronavirus structure reveals targets for vaccines and treatments. Retrieved February 25, 2022, from www.nih.gov/ne ws-events/nih-research-matters/novel-coronavirus-structure-reveals-targets-va ccines-treatments

Nelkin, D. (1995). *Selling Science: How the Press Covers Science and Technology*. Revised edition. New York: W. H. Freeman and Company.

Overney, N. & Overney, G. (2011). The history of photomicrography. Retrieved February 11, 2022, from www.microscopy-uk.org.uk/mag/artmar10/history_ photomicrography_ed3.pdf

Pennisi, E. (1997). Haeckel's embryos: Fraud rediscovered. *Science*, *277*(5331): 1435.

Richards, R. J. (2008). Haeckel's embryos: Fraud not proven. *Biology & Philosophy*, *24*(1), 147–154.

Richardson, M. K., Hanken, J., Gooneratne, M. L., et al. (1997). There is no highly conserved embryonic stage in the vertebrates: Implications for current theories of evolution and development. *Anatomy and Embryology*, *196*(2), 91–106.

Troemel, E. R., Félix, M.-A., Whiteman, N. K., Barrière, A., & Ausubel F. M. (2008). Microsporidia are natural intracellular parasites of the nematode Caenorhabditis elegans. *PLoS Biology*, 6(12), e309

Watts, E., Levit, G. S., & Hossfeld, U. (2019). Ernst Haeckel's contribution to Evo-Devo and scientific debate: A re-evaluation of Haeckel's controversial illustrations in US textbooks in response to creationist accusations. *Theory in Biosciences*, *138*(1), 9–29.

Zhou, P., Yang, X.-L., Wang, X.-G., et al. (2020). A pneumonia outbreak associated with a new coronavirus of probable bat origin. *Nature*, *579*(7798), 270–273. Cells that expressed mouse ACE2 were an exception and didn't get infected.

## Chapter 4

BBVA Foundation Department of Social Studies and Public Opinion (2017). BBVA foundation international study on scientific culture. Retrieved March 18, 2022, from www.fbbva.es/wp-content/uploads/2017/05/dat/Understandingsciencen otalarga.pdf

Christiansen, J. (2013). A defense of artistic license in illustrating scientific concepts for a non-specialist audience. In *Communicating Complexity 2013 Conference Proceedings* (pp. 49–60). Rome: Edizioni Nuova Cultura-Roma.

Giaimo, C. (2020). The spiky blob seen around the world. *The New York Times*. Retrieved March 18, 2022, from www.nytimes.com/2020/04/01/health/corona virus-illustration-cdc.html

Gross, A. G., & Harmon, J. E. (2013). *Science from Sight to Insight: How Scientists Illustrate Meaning*. Chicago, IL: University of Chicago Press.

Harp, S. F. & Mayer, R. E. (1998). How seductive details do their damage: A theory of cognitive interest in science learning. *Journal of Educational Psychology*, *90*(3), 414–434.

Lakoff, G. & Johnson, M. (2003). *Metaphors We Live By* (2nd ed.). Chicago, IL: University of Chicago Press.

McDougall, S. J., de Bruijn, O., & Curry, M. B. (2000). Exploring the effects of icon characteristics on user performance: The role of icon concreteness, complexity, and distinctiveness. *Journal of Experimental Psychology: Applied, 6*(4), 291–306.

Middleton, A., Milne, R., Almarri, M. A., et al. (2020). Global public perceptions of genomic data sharing: What shapes the willingness to donate DNA and health data? *The American Journal of Human Genetics, 107*(4), 743–752.

Phillips, B. J., & McQuarrie, E. F. (2004). Beyond visual metaphor: A new typology of visual rhetoric in advertising. *Marketing Theory, 4*(1), 113–136.

Reynolds, A. S. (2022). *Understanding Metaphors in the Life Sciences*. Cambridge: Cambridge University Press.

Taylor, C. & Dewsbury, B. M. (2018). On the problem and promise of metaphor use in science and science communication. *Journal of Microbiology & Biology Education, 19*(1), 1–5.

## Chapter 5

Burke, C. (2009). Isotype: Representing social facts pictorially. *Information Design Journal, 17*(3), 211–223.

Cairo, A. (2015). Graphics lies, misleading visuals: Reflections on the challenges and pitfalls of evidence-driven visual communication. In D. Bihanic (Ed.), *New Challenges for Data Design* (pp. 103–116). London: Springer.

Cleveland, W. & McGill, R. (1984). Graphical perception: Theory, experimentation, and application to the development of graphical methods. *Journal of the American Statistical Association, 79*(387), 531–554.

Fagerlin, A., Wang, C., & Ubel, P. A. (2005). Reducing the influence of anecdotal reasoning on people's health care decisions: Is a picture worth a thousand statistics? *Medical Decision Making: An International Journal of the Society for Medical Decision Making, 25*(4), 398–405.

Friendly, M. & Denis, D. (2005). The early origins and development of the scatterplot. *Journal of the History of the Behavioral Sciences, 41*(2), 103–130.

Garcia-Retamero, R., Galesic, M., & Gigerenzer, G. (2010). Do icon arrays help reduce denominator neglect? *Medical Decision Making, 30*(6), 672–684.

Kosara, R. (2016). Stacked bars are the worst. Retrieved November 15, 2023, from https://eagereyes.org/techniques/stacked-bars-are-the-worst

Mitropoulos, A., Brownstein, J., and Bhatt, J. (2021). COVID-19 vaccinations finally starting to stem pandemic's tide in US: Analysis. Retrieved April 7, 2022, from https://abcnews.go.com/Health/covid-19-vaccinations-finally-starting-stem-pandemics-tide/story?id=77514759

Pew Research Center (2015). American trends panel wave 6. Retrieved November 15, 2023, from www.pewresearch.org/fact-tank/2015/09/16/the-art-and-science-of-the-scatterplot

Playfair, W. (1801). *The Statistical Breviary*. London: T. Bensley.

Romano, A., Sotis, C., Dominioni, G., & Guidi, S. (2020). The scale of COVID-19 graphs affects understanding, attitudes, and policy preferences. *Health Economics*, *29*(11), 1482–1494.

Ryan, W. H. & Evers, E. R. K. (2020). Graphs with logarithmic axes distort lay judgments. *Behavioral Science & Policy*, *6*(2), 13–23.

Talbot, J., Setlur, V., & Anand, A. (2014). Four experiments on the perception of bar charts. *IEEE Transactions on Visualization and Computer Graphics*, *20*(12), 2152–2160.

Tufte, E. (2001). *The Visual Display of Quantitative Information*. Cheshire, CT: Graphics Press LLC.

Yang, B. W., Restrepo, C. V., Stanley, M. L., & Marsh, E. J. (2021). Truncating bar graphs persistently misleads viewers. *Journal of Applied Research in Memory and Cognition*, *10*(2), 298–311.

## Chapter 6

Centers for Disease Control and Prevention (2024). COVID-19 vaccinations by county. Retrieved November 20, 2023, from https://covid.cdc.gov/covid-data-tracker/#county-view

Cooper, S., Khatib, F., Treuille, A., et al. (2010). Predicting protein structures with a multiplayer online game. *Nature*, *466*(7307), 756–760. Supplementary information. Retrieved November 22, 2023, from https://static-content.springer.co

m/esm/art%3A10.1038%2Fnature09304/MediaObjects/41586_2010_BFnatur
e09304_MOESM302_ESM.pdf

Cooper, S., Treuille, A., Barbero, J., et al. (2010). The challenge of designing scientific discovery games. In *Proceedings of the Fifth International Conference on the Foundations of Digital Games*, 40–47.

Daston, L. & Galison, P. (2007). *Objectivity*. New York: Zone Books.

Grossman, L. (2010). Computer game makes you a genetic scientist. *WIRED*. Retrieved May 5, 2022, from www.wired.com/2010/11/phylo-game

Hirst, J. D., Glowacki, D. R., & Baaden, M. (2014). Molecular simulations and visualization: Introduction and overview. *Faraday Discussions*, *169*, 9–22.

Johns Hopkins Coronavirus Resource Center (2022). The demographics of COVID. Retrieved November 20, 2023, from https://coronavirus.jhu.edu/data/racial-d ata-transparency

Johns Hopkins Coronavirus Resource Center (2023). Cumulative cases over time. Retrieved November 20, 2023, from https://coronavirus.jhu.edu/data/ani mated-world-map

Kawrykow, A., Roumanis, G., Kam, A., et al. (2012). Phylo: A citizen science approach for improving multiple sequence alignment. *PLoS One*, *7*(3), e31362.

Khatib, F., DiMaio, F., Foldit Contenders Group, et al. (2011). Crystal structure of a monomeric retroviral protease solved by protein folding game players. *Nature Structural & Molecular Biology*, *18*(10), 1175–1177.

von Ahn, L. (2006). Games with a purpose. *Computer*, 96–98.

# Chapter 7

Baxter, M., Lonsdale, M., & Westland, S. (2021). Utilising design principles to improve the perception and effectiveness of public health infographics. *Information Design Journal*, *26*(2), 124–156.

Edwards, J. L. (1997). *Political Cartoons in the 1988 Presidential Campaign: Image, Metaphor, and Narrative*. New York: Garland Publishing.

Krum, R. (2013). *Cool Infographics: Effective Communication with Data Visualization and Design*. Indianapolis, IN: Wiley.

Lonsdale, M. & Lonsdale, D. (2019). Information visualisation. Retrieved May 26, 2022, from www.researchgate.net/profile/Maria-Lonsdale/publication/330825479_Design2Inform_Information_visualisation/links/5e32c86a458515072d6f3ee7/Design2Inform-Information-visualisation.pdf

Miller, G. A. (1956). The magical number seven, plus or minus two: Some limits on our capacity for processing information. *Psychological Review*, *63*(2), 81–97.

Polman J. & Gebre, E. (2015). Towards critical appraisal of infographics as scientific inscriptions. *Journal of Research in Science Teaching*, *52*(6), 868–893.

Stones, C. & Gent, M. (2015). The 7 G.R.A.P.H.I.C. principles of public health infographic design. Retrieved May 26, 2022, from https://visualisinghealth.files.wordpress.com/2014/12/guidelines.pdf

# Index